Universitext

Universitext

Editors: F. W. Gehring, P.R. Halmos, C.C. Moore

Chern: Complex Manifolds Without Potential Theory
Chorin/Marsden: A Mathematical Introduction to Fluid Mechanics
Cohn: A Classical Invitation to Algebraic Numbers and Class Fields
Curtis: Matrix Groups
van Dalen: Logic and Structure
Devlin: Fundamentals of Contemporary Set Theory
Edwards: A Formal Background to Mathematics I a/b
Edwards: A Formal Background to Higher Mathematics II a/b
Endler: Valuation Theory
Frauenthal: Mathematical Modeling in Epidemiology
Gardiner: A First Course in Group Theory
Godbillon: Dynamical Systems on Surfaces
Greub: Multilinear Algebra
Hermes: Introduction to Mathematical Logic
Kalbfleish: Probability and Statistical Inférence I/II
Kelly/Matthews: The Non-Euclidian, The Hyperbolic Plane
Kostrikin: Introduction to Algebra
Luecking/Rubel: Complex Analysis: A Functional Analysis Approach
Lu: Singularity Theory and an Introduction to Catastrophe Theory
Marcus: Number Fields
Meyer: Essential Mathematics for Applied Fields
Moise: Introductory Problem Course in Analysis and Topology
Rees: Notes on Geometry
Reisel: Elementary Theory of Metric Spaces
Rey: Introduction to Robust and Quasi-Robust Statistical Methods (in prep)
Rickart: Natural Function Algebras
Schreiber: Differential Forms
Tolle: Optimization Methods

D. H. Luecking
L. A. Rubel

Complex Analysis
A Functional Analysis Approach

Springer-Verlag
New York Berlin Heidelberg Tokyo

D. H. Luecking
University of Arkansas
Department of Mathematics
Fayetteville, AR 72701
U.S.A.

L. A. Rubel
University of Illinois
Department of Mathematics
Urbana-Champaign, IL 61801
U.S.A.

AMS Classification: 30-01, 30G15, 46N05

Library of Congress Cataloging in Publication Data
Luecking, Daniel H.
Complex analysis.
(Universitext)
Bibliography: p.
1. Functions of complex variables. 2. Functional analysis. I. Rubel, Lee A. II. Title.
QA331.L818 1984 515.9 84-5354

With 7 Illustrations.

Printed and bound by R.R. Donnelley & Sons, Harrisonburg, Virginia.
Printed in the United States of America.

9 8 7 6 5 4 3 2 1

ISBN 0-387-90993-1 Springer-Verlag New York Berlin Heidelberg Tokyo
ISBN 3-540-90993-1 Springer-Verlag Berlin Heidelberg New York Tokyo

Introduction

The main idea of this book is to present a good portion of the standard material on functions of a complex variable, as well as some new material, from the point of view of functional analysis. The main object of study is the algebra $H(G)$ of all holomorphic functions on the open set G, with the topology on $H(G)$ of uniform convergence on compact subsets of G.

From this point of view, the main theorem of the theory is Theorem 9.5, which concretely identifies the dual of $H(G)$ with the space of germs of holomorphic functions on the complement of G. From this result, for example, Runge's approximation theorem and the global Cauchy integral theorem follow in a few short steps. Other consequences of this duality theorem are the Germay interpolation theorem and the Mittag-Leffler Theorem. The approach via duality is entirely consistent with Cauchy's approach to complex variables, since curvilinear integrals are typical examples of linear functionals.

The prerequisite for the book is a one-semester course in complex variables at the undergraduate-graduate level, so that the elements of the local theory are supposed known. In particular, the Cauchy Theorem for the square and the circle are assumed, but not the global Cauchy Theorem in any of its forms. The second author has three times taught a graduate course based on this material at the University of Illinois, with good results.

Our experience in teaching the course is that the learning curve is rather flat in the beginning, but then takes off sharply about midway through the course. The reason is that in the beginning there are a lot of preliminaries to get out of the way, but once the student gets these under his belt they really pay off. The learning curve turns sharply up and continues that way.

Many of the students wondered at first if this course would prepare them for the Preliminary Examination Unit in complex analysis. Comparing the material actually covered in Spring 1983 with the standard syllabus, we find that all standard topics were covered with the exception of harmonic functions and the Poisson integeral formula. Instructors could easily incorporate these and other topics into their courses while retaining the spirit of our approach.

The authors acknowledge an inner conflict. On the one hand, we feel that "the subject is being done right", while on the other hand, we regret the omission of some of the beautiful and important classical material--the Fabry Gap Theorem, the Euler Gamma Function, or Elliptic Functions, for example. But these are sacrifices made for consistency and generality of method and viewpoint as well as establishing vital contact with the rest of the graduate mathematics curriculum.

Beginning in Chapter 8, some of the exercises are starred to indicate that we have not worked them out. Most of the starred exercises are probably more difficult than most of the unstarred ones.

Contents

§1. Preliminaries: Set Theory and Topology

We assume familiarity with the rudiments of informal set theory including such notions as set, subset, superset, the null set $\emptyset$, the union or intersection of a family of sets, set difference $(A\setminus B)$, complement (compl A), Cartesian product; functions, domain, range, one-to-one, onto, image, inverse, restriction; partial ordering, linear (or total) ordering, and equivalence relation.

We require some form of "Zorn's Lemma" and we choose the following.

(1.1) Hausdorff Maximality Principal. In a partially ordered set, any linearly ordered subset is contained in a maximal linearly order set.

(1.2) Definition. A topology on a set X is a collection Ω of subsets of X which contains both $\emptyset$ and X and is closed under arbitrary unions and finite intersections.

More precisely (i) $\emptyset \in \Omega$, $X \in \Omega$, (ii) if $\Omega_1 \subset \Omega$ then $\bigcup_{\omega\in\Omega_1} \omega \in \Omega$, (iii) if $\omega_1,\ldots,\omega_n$ belong to Ω then $\omega_1 \cap \omega_2 \cap\cdots\cap \omega_n \in \Omega$.

When X is assigned a topology Ω we call X a topological space and the elements of Ω are called open sets. A set $\lambda \subseteq X$ is called closed if its complement $X\setminus\lambda$ is open. It is a trivial exercise to verify that any intersection of closed sets is closed, and that a finite union of closed sets is closed.

(1.3) Definition. If Ω is any collection of subsets of X, we define Ω^*, the topology generated by Ω, to be the smallest topology containing Ω. Ω is then said to be a sub-basis for Ω^*.

The "smallest topology containing Ω" means the intersection of all topologies that contain Ω. Since the collection of all subsets of X is a topology, this intersection makes sense (i.e. there are such topologies). It is clear that Ω^* must contain all unions of finite interections of members of Ω, and since the family of all such sets is a topology, Ω^* contains only these. If Ω is closed under finite interections (so that Ω^* consists of arbitrary unions of elements of Ω) then Ω is said to be a basis for Ω^*.

We will frequently use the somewhat loose terminology "X is a topological space" to mean that some fixed but unspecified topology on

X is understood.

(1.4) Definition. (Continuous) If X and X' are topological spaces and $\varphi : X \to X'$ is a function, then φ is called continuous if $\varphi^{-1}(\omega')$ is open in X whenever ω' is open in X'. (More precisely, if Ω and Ω' are the topologies on X and X' respectively, then we require $\varphi^{-1}(\omega') \in \Omega$ whenever $\omega' \in \Omega'$.) (Homeomorphism) If φ is one-to-one and onto and both φ and φ^{-1} are continuous, φ is called a homeomorphism. If a homeomorphism of X onto X' exists then X and X' are called homeomorphic spaces and we write $X \sim X'$ (or $(X,\Omega) \sim (X',\Omega')$ if the topologies need to be emphasized.)

If $\varphi : X \to X'$ is a homeomorphism then a set $\omega \subset X$ is open if and only if $\varphi(\omega) \subset X'$ is open. Since one-to-one and onto functions preserve all set theoretic operations, any property of (X,Ω) expressed in set theory must be shared by (X',Ω'). Any property a topological space may have which must be shared by all homeomorphic spaces is called a topological property.

It is easy to verify that a composition of continuous functions is continuous and hence that $\sim$ is an equivalence relation. It is frequently convenient not to distinguish two spaces if they are homeomorphic.

The properties expressed in the following definition are topological properties.

(1.5) Definition. (Hausdorff) X is said to be a Hausdorff space if for any two points $x, x' \in X$ with $x \neq x'$, there exist disjoint open sets ω and ω' with $x \in \omega$, $x' \in \omega'$. (Connected) To say that X is connected is to say that the only subsets of X which are simultaneously open and closed are $\emptyset$ and X.

A set is called clopen if it is both open and closed.

(1.6) Definition. If (X,Ω) is a topological space and $Y \subset X$, the relative topology on Y induced by Ω is the family $\Omega' = \{\omega \cap Y : \omega \in \Omega\}$. We write $(Y;X,\Omega)$ to denote the topological space (Y,Ω').

To say that a subset $Y \subset X$ has a topological property is to say that $(Y;X,\Omega)$ has this property.

(1.7) Chaining Lemma. If $\{Y_\alpha\}$ is a collection of connected subsets of X and if $\bigcap_\alpha Y_\alpha \neq \emptyset$, then $\bigcup_\alpha Y_\alpha$ is connected.

Proof. Let $C = \bigcup_\alpha Y_\alpha$ and suppose A is a clopen set in C. We may suppose $A \cap \bigcap_\alpha Y_\alpha \neq \emptyset$ (otherwise replace A by $C \setminus A$ which is also clopen in C). Then each set $A \cap Y_\alpha$ is a non-empty clopen set in Y_α. Since Y_α is connected $A \cap Y_\alpha = Y_\alpha$. This means $Y_\alpha \subset A$ for all α, so that $C = A$. That is, the only non-empty clopen set in C is C itself, so C is connected. QED

Note that C itself need not be a clopen set in X. We are saying that, as a topological space in its own right, C is the only clopen subset of C (save $\emptyset$).

(1.8) Corollary. If Y is a connected subset of X and $\{Y_\alpha\}$ is a collection of connected subsets of X such that $Y \cap Y_\alpha \neq \emptyset$ for all α, then $Y \cup \bigcup_\alpha Y_\alpha$ is connected.

Proof. By (1.7) each $Y \cup Y_\alpha$ is connected and again by (1.7) $\bigcup_\alpha (Y \cup Y_\alpha) = Y \cup \bigcup_\alpha Y_\alpha$ is connected. QED

(1.9) Proposition. If X is connected and $f : X \to Y$ is continuous, then $f(X)$ is connected.

Proof. Exercise.

(1.10) Definition. If $x \in X$, then the union of all connected sets containing x is called the connected component of X containing x or simply the component of X containing x.

It is clear that every $x \in X$ is contained in some connected set (namely $\{x\}$) so the component is well defined. It is also clear from the chaining lemma that the component is indeed connected. Also from the chaining lemma it follows that any component of X containing x is either disjoint from or equal to the component containing x' for any other $x' \in X$. Thus the various components of X partition X into disjoint sets.

(1.11) Definitions. (Closure) If $Y \subset X$ then the closure of Y,

denoted Y^- is the intersection of all closed supersets of Y.
(Boundary) The boundary of Y is the set $\partial Y = Y^- \cap (\text{compl } Y)^-$.
(Dense) Y is dense in X means $Y^- = X$.
(Separable) X is separable if it contains a dense, countable subset.

Clearly Y^- and ∂Y are closed sets. X is alway dense in itself and so any countable space X is separable.

(1.12) More Definitions. (Open Cover) If $Y \subset X$, then a set Ω' of open sets is called an open cover of Y if $Y \subset \bigcup_{\omega \in \Omega'} \omega$.
(Compact) X is compact means that every open cover of X contains a finite subcover (i.e. a finite subcollection which is also an open cover).
(Baire Space) X is a Baire Space if every countable family of dense open sets has a dense intersection .
(Neighborhood) If $x \in \omega \subset A$, where ω is open, then A is called a neighborhood of x. If Y is a set with $Y \subset \omega \subset A$, then A is a neighborhood of Y.

When the set of real numbers $\mathbb{R}$ or the set of complex numbers $\mathbb{C}$ is considered as a topological space it is assumed to have the usual topology. This is the topology generated by the following subbasis:

$$\Omega = \{B(x,\varepsilon) : x \in \mathbb{R} \ (\text{or } \mathbb{C}),\ \varepsilon > 0\}$$

where

$$B(x,\varepsilon) = \{y \in \mathbb{R} \ (\text{or } \mathbb{C}) : |x - y| < \varepsilon\}.$$

In the case of $\mathbb{R}$, Ω is the collection of open intervals and in $\mathbb{C}$ it consists of open disks. It is clear that the usual topology on $\mathbb{R}$ is the same as the relative topology it inherits as a subset of $\mathbb{C}$.

"Continuous" has its usual meaning for complex-valued functions on $\mathbb{R}$ or $\mathbb{C}$. $\mathbb{R}$ and $\mathbb{C}$ are both connected but $\mathbb{R}\setminus\{0\}$ is not connected. (This shows that $\mathbb{R}$ and $\mathbb{C}$ are not homeomorphic since the removal of a point from $\mathbb{C}$ leaves it connected.) The closure of an open interval in $\mathbb{R}$ is the corresponding closed interval. An example of a compact set is $[0,1]$. It is also a Baire space. Its compactness follows from the Heine-Borel Theorem. That it is a Baire space is a consequence of the Baire Category Theorem, to be proved later.

(1.13) Definition. A curve γ in X is a continuous functions $\gamma : [0,1] \to X$. The associated physical curve $\gamma^{\wedge}$ is the range of γ. We say γ is a closed curve if $\gamma(0) = \gamma(1)$ and that it is a simple closed curve if $\gamma(a) \neq \gamma(b)$ whenever $a \neq b$ unless $\{a,b\} = \{0,1\}$. Two curves γ and γ' are equivalent, written $\gamma \sim \gamma'$, if there is a strictly increasing continuous function $\lambda : [0,1] \to [0,1]$ such that $\lambda(0) = 0$ and $\lambda(1) = 1$ and $\gamma' = \gamma \circ \lambda$.

The fact that $[0,1]$ is connected and compact implies that any physical curve is connected and compact. The proofs are exercises.

(1.14) Definition. (Interior) If A is a subset of the topological space X, the interior of A, written int A, is the union of all open subsets of A (open in X).

(Nowhere dense) If $\text{int}(A^{-}) = \emptyset$ then A is nowhere dense in X.

(First category) A is said to be of first category in X if it is a union of countably many nowhere dense sets ("meager in X" is also used).

(Second category) A is of second category in X if it is not of first category.

Important note: all of these concepts depend on the ambient space X. For instance, if A is considered as a topological space in its own right then int A = A since (with the relative topology) A is open. Thus, if the space X is not made unambiguous, such phrasings as "A is of second category in X" should be used

(1.15) Proposition. X is a Baire space if and only if every non-empty open subset of X is of second category in X.

Proof. This amounts to replacing intersections of open sets by unions of closed sets via DeMorgan's laws. Trivial. QED

This proposition is frequently used to show that a given set is non-empty by showing that its complement is of first category and that the ambient space is a Baire space. This becomes most useful when we have a lot of examples of Baire spaces. This will be the case after the Baire Category Theorem (1.19).

(1.16) Definition. Given a set X, a metric on X is a non-negative function ρ on $X \times X$ with the following properties

a.) $\rho(x,y) = 0$ if and only if $x = y$.

b.) $\rho(x,y) = \rho(y,x)$, all $x, y \in X$.

c.) $\rho(x,y) \leq \rho(x,z) + \rho(z,y)$, all $x, y, z \in X$.

A pair (X,ρ) where ρ is a metric on X is called a metric space. The metric topology on X is the topology generated by sets of the form $B(x,\varepsilon) = \{y \in X : \rho(x,y) < \varepsilon\}$.

If (X,Ω) is a topological space which is homeomorphic to a space with a metric topology, we will say (X,Ω) is metrizable (or even call it a metric space). All metric spaces are Hausdorff. Both the complex plane $\mathbb{C}$ and the real line $\mathbb{R}$ with their usual topology are metric spaces $(\rho(x,y) = |x - y|)$. Another metric on $\mathbb{C}$ that gives the same topology is the chordal metric defined by

$$\rho_c(a,b) = \frac{|a - b|}{\sqrt{1 + |a|^2}\,\sqrt{1 + |b|^2}}.$$

The chordal metric is the distance between the points on the Riemann sphere that project onto a and b under stereographic projection. (More precisely, view $\mathbb{C}$ as the plane $x_3 = 0$ in $\mathbb{R}^3$ and let S be the sphere $x_1^2 + x_2^2 + (x_3 - \frac{1}{2})^2 = \frac{1}{4}$. For $z' \in S$, let $P(z')$ be the intersection with $\mathbb{C}$ of the line through $(0,0,1)$ and z'. Then $\rho_c(a,b)$ is the (Euclidean) distance in $\mathbb{R}^3$ between $P^{-1}(a)$ and $P^{-1}(b)$.) Note that ρ_c can be extended to $(\mathbb{C} \cup \{\infty\}) \times (\mathbb{C} \cup \{\infty\})$ by $\rho_c(a,\infty) = \frac{1}{\sqrt{1 + |a|^2}}$ and $\rho_c(\infty,\infty) = 0$. $\mathbb{C} \cup \{\infty\}$ with this topology will be denoted $\mathbb{C}^\wedge$.

(1.17) Definitions. (Cauchy sequence) In a metric space (X,ρ) a sequence $\{x_n\}$ is a Cauchy sequence provided that for each $\varepsilon > 0$, there exists $n(\varepsilon)$ such that $m, n > n(\varepsilon)$ imply $\rho(x_n,x_m) < \varepsilon$. (Convergence) In a Hausdorff space, a sequence $\{x_n\}$ converges to x $(x_n \to x)$ if for each neighborhood U of x there is an n_0 such that $x_n \in U$ for all $n > n_0$. We say $\{x_n\}$ converges if such an x exists.

There is no notion of Cauchy sequence for the general topological space. It is easy to show that, in a metric space, $x_n \to x$ if and only if $\rho(x_n, x) \to 0$.

(1.18) Definition. A metric space is called complete if every Cauchy sequence converges.

(1.19) Baire Category Theorem. A complete metric space X is a Baire space.

Proof. Let U_n, $n = 1,2,3,\ldots$ be dense open sets. We must show $\cap_n U_n$ is dense. For this it suffices to show that $B(x_0,\varepsilon) \cap \cap_n U_n \neq \emptyset$ for any $x_0 \in X$ and $\varepsilon > 0$. Since $B(x_0,\varepsilon) \cap U_1$ is a non-empty open set there exists a ball $B(x_1,r_1)^- \subseteq B(x_0,\varepsilon) \cap U_1$, with $\rho(x_0,x_1) < \varepsilon/4$ and $r_1 < \varepsilon/4$. Now $B(x_1,r_1) \cap U_1 \cap U_2$ is non-empty so there exists $B(x_2,r_2)$ with $B(x_2,r_2)^- \subset B(x_1,r_1) \cap U_1 \cap U_2$ and $r_2 < r_1/2$. In general, there exist $B(x_n,r_n)$ for $n = 1,2,3,\ldots$ such that

$$B(x_n,r_n)^- \subset B(x_{n-1},r_{n-1}) \cap \bigcap_{k=1}^{n-1} U_k$$

and $r_n < r_1/2^{n-1}$. Now for $m \geq n$, $\rho(x_n,x_m) \leq \rho(x_n,x_{n+1}) + \cdots + \rho(x_{m-1},x_m) \leq 2^{-n}\varepsilon$ so the sequence $\{x_n\}$ is a Cauchy sequence. Let x be its limit. Since $x_n \in B(x_k,r_k)$ for $n > k$ it follows that the limit x belongs to $B(x_k,r_k)^-$ for all k and thus $x \in U_k$ for all k. Clearly also $x \in B(x_0,\varepsilon)$. QED

In this proof we used the fact that $[x_n \in A, x_n \to x] \Rightarrow x \in A^-$. This is not at all hard to prove and follows, in any case from (1.22b) below.

(1.20) Definitions. (Distance to a set) If (X,ρ) is a metric space and $A \subset X$, then $\rho(x,A) = \inf\{\rho(x,y) : y \in A\}$. If also $B \subset X$ then

$$\rho(A,B) = \inf\{\rho(x,y) : x \in A,\ y \in B\}.$$

(Diameter) The diameter $\delta(A)$ of A is given by $\delta(A) = \sup\{\rho(x,y) : x \in A,\ y \in A\}$.

When ρ is unambiguous or unspecified $\mathrm{dist}(x,A)$ and $\mathrm{dist}(A,B)$ may be used in place of $\rho(x,A)$ and $\rho(A,B)$.

Next we translate the definitions of some topological concepts into their corresponding definitions in terms of neighborhoods.

(1.21) **Proposition**. *In a topological space* X

a.) *A set is open if and only if it contains a neighborhood of each of its points.*

b.) *A point* x *is in the closure of a set* A *if and only if every neighborhood of* x *intersects* A.

c.) $x \in \operatorname{int} A$ *if and only if there is a neighborhood* N_x *of* x *with* $N_x \subset A$.

d.) A *is dense if and only if for all* $x \in X$ *and all neighborhoods* N *of* x *we have* $N \cap A \neq \emptyset$.

There is a similar translation into metric terms.

(1.22) **Proposition**: *In a metric space* (X,ρ)

a.) *A set* U *is open if and only if for every* $x \in U$ *there is an* $\varepsilon > 0$ *such that* $B(x,\varepsilon) \in U$.

b.) *A point* x *is in the closure of* A *if and only if* $\rho(x,A) = 0$, *and this is true if and only if there is a sequence* $\{x_n\}$ *in* A *with* $x_n \to x$.

c.) $x \in \operatorname{int} A$ *if and only if there is an* $\varepsilon > 0$ *with* $B(x,\varepsilon) \subset A$.

d.) A *is dense in* X *if and only if for all* $x \in X$ *and* $\varepsilon > 0$, *we have* $B(x,\varepsilon) \cap A \neq \emptyset$.

We remark that Definition (1.16)(c) implies $|\rho(x,y) - \rho(x',y')| < \rho(x,x') + \rho(y,y')$, which easily implies the continuity of ρ on $X \times X$ (provided we define the topology on $X \times X$ as in (1.23) below). It is also an easy exercise to show that $\rho(x,A)$ is continuous in x for each $A \subset X$.

(1.23) **Definition**. *If* $(X_1,\Omega_1),\ldots,(X_n,\Omega_n)$ *are topological spaces, then the* **product topology** *on* $X_1 \times X_2 \times \cdots \times X_n$ *is the topology generated by* $\{\omega_1 \times \cdots \times \omega_n : \omega_i \in \Omega_i,\ i = 1,2,\ldots,n\}$.

We end this chapter with a collection of miscellaneous results about continuous functions and compact sets.

(1.24) **Proposition**. *Any compact subset of a Hausdorff space is closed. Any closed subset of a compact set is compact.*

Proof. Suppose A is compact and $p \in \text{compl } A$. For each $q \in A$ there is a neighborhood N_q of q and a neighborhood N'_q of p, such that $N_q \cap N'_q = \emptyset$. The collection $\{N_q : q \in A\}$ covers A and so has a finite subcover $N_{q_1}, \ldots, N_{q_k}$. Then $N'_{q_1} \cap N'_{q_2} \cap \cdots \cap N'_{q_k}$ is a neighborhood of p that does not intersect A. Thus $\text{compl } A$ is open.

If A is a closed subset of a compact set X and if $\{U_\alpha\}$ is an open cover of A, then $\{U_\alpha\} \cup \{X \setminus A\}$ is an open covering of X. It therefore has a finite subcover: $U_{\alpha_1}, \ldots, U_{\alpha_n}, X \setminus A$. Then $U_{\alpha_1}, \ldots, U_{\alpha_n}$ is a finite covering of A. QED.

(1.25) Proposition. If X is compact and $f : X \to \mathbb{R}$ (or $\mathbb{C}$) is continuous, then f is bounded.

Proof. (For $\mathbb{R}$) $\{U_n = f^{-1}((-n,n)) : n = 1,2,\ldots\}$ is an open cover of X and so finitely many sets $U_1, \ldots, U_N$ cover X. Clearly $f(X) \subset (-N,N)$ and so is bounded. The same sort of proof works for $\mathbb{C}$. QED.

(1.26) Proposition. If X is compact and Y is a topological space and $f : X \to Y$ is continuous, then $f(X)$ is compact.

Proof. Exercise.

(1.27) Definition. If $\{x_n\}$ is a sequence in X and $x \in X$, then x is called a limit point of $\{x_n\}$ provided every neighborhood of x contains x_n for infinitely many n.

(1.28) Proposition. If A is a compact metric space, then any sequence in A has a convergent subsequence.

Proof. We show that if $\{x_n\}$ is in A, then $\{x_n\}$ has a limit point. For if it has no limit point, then for every $p \in A$ there is a neighborhood N_p containing x_n for only finitely many n. A finite subcover of $\{N_p : p \in A\}$ shows that A contains x_n for only finitely many n, a contradiction. Thus a limit point, x, exists. Choose $x_{n_k} \in B(x, \frac{1}{k})$. Then $x_{n_k} \to x$. QED

(1.29) Proposition. Suppose every sequence in a metric space X has a convergent subsequence. Then X is compact.

Proof. Let $\{U_i\}$ be an open cover. First we claim there exists an $\varepsilon > 0$ such that for any $x, B(x,\varepsilon)$ is contained in some U_i. If not, there exists a sequence $\{x_n\}$ (which we may suppose to be convergent) and $\varepsilon_n \to 0$ such that $B(x_n,\varepsilon_n)$ is not contained in any U_i. Let $x_n \to x$. Then $x \in U_i$ for some i, and so $B(x,2\varepsilon) \subset U_i$ for some i and for some $\varepsilon > 0$. If $x_n \in B(x,\varepsilon)$ and $\varepsilon_n < \varepsilon$, then $B(x_n,\varepsilon_n) \subset U_i$, a contradiction. Now we claim that $\{B(x,\varepsilon):x \in X\}$ has a finite subcover. If not, choose x_1, $B_1 = B(x_1,\varepsilon)$, then choose $x_2 \notin B_1$ and $B_2 = B(x_2,\varepsilon)$. Once $x_1,\ldots,x_n$ have been chosen, choose $x_{n+1} \notin B_1 \cup\cdots\cup B_n$. The resulting sequence $\{x_n\}$ satisfies $\rho(x_n,x_m) \geq \varepsilon$ for all n, m and so $\{x_n\}$ cannot converge. Thus $\{B(x,\varepsilon): x \in X\}$ has a finite subcover, and the corresponding U_i form a finite cover of X. QED

(1.30) Proposition: Any subspace of a separable metric space is separable.

Proof. Let X be a metric space with $\{x_1,x_2,x_3,\ldots\}$ a dense subset. Let $Y \subset X$. Construct a countable subset of Y by choosing one point from $Y \cap B(x_n,1/m)$ whenever this intersection is non-empty. The resulting set is dense in Y. QED

NOTES: The definitions and proofs here are standard (at least as standard as such things get). There are numerous excellent texts on these topics. Two well-known ones are [J. L. Kelley] and [J. Dugundji].

Exercises

1. Prove that removal of a point from $\mathbb{C}$ leaves $\mathbb{C}$ connected. For the more ambitious: prove that removal of countably many points leaves $\mathbb{C}$ connected.

2. Suppose $f : \mathbb{C} \to \mathbb{C}$ and $\lim_{\substack{z\to z_0 \\ z\neq z_0}} f(z) = 0$ for every $z_0 \in \mathbb{C}$. Prove that f has at least one zero.

3. A topological space X is said to be path-connected if any two points $x, y \in X$ can be connected by a curve in X. Show that any open set in $\mathbb{C}$ is connected if and only if it is path-connected. More generally, show that X is path-connected provided it is connected and every point has a path-connected neighborhood.

4. Show that a bounded closed set $F \subseteq \mathbb{C}$ is connected iff for every pair of points $z, z' \in F$ and every $\varepsilon > 0$ there exists a positive integer n and points $z_1, z_2, \ldots, z_n$ in F such that $|z - z_1| < \varepsilon$, $|z_n - z'| < \varepsilon$ and $|z_i - z_{i+1}| < \varepsilon$, $i = 1,2,\ldots,n-1$.

5. Show that the closure of a connected set is connected. Show that any connected component of a topological space is closed but need not be open. Show that they are open if and only if every point has a connected neighborhood.

6. A Hausdorff space X is locally compact if for every $x \in X$ and every neighborhood N of x, there is a neighborhood U of x with $U^- \subseteq N$ and U^- compact. Show that any locally compact Hausdorff space is a Baire space. (Hint: Replace the shrinking sequence of balls in (1.19) by a shrinking sequence of compact sets.)

7. In (a) through (f) find a set $A \subseteq \mathbb{C}$ having the stated property.

a) A is open and $\operatorname{int}(A^-) \neq A$.

b) A is compact and connected and $\operatorname{int} A$ has infinitely many components.

c) $A = A_1 \cup A_2$, A_1 and A_2 are open, disjoint and non-empty, $\partial A_1 = \partial A_2 = \partial A$.

d) A is closed, uncountable and totally disconnected. (i.e., each component of A consists of exactly one point.)

e) A is open and $\operatorname{compl} A$ is connected but $\operatorname{compl}(A^-)$ is not connected.

f) A is open and $\operatorname{compl} A$ has uncountably many components.

8. If X is a topological space and $x \in X$, then x is called an isolated point if $x \notin (\operatorname{compl}\{x\})^-$. $X \neq \emptyset$ is said to be perfect if it contains no isolated points. Show that a perfect compact space must be uncountable. (Find disjoint, non-empty perfect closed subspaces A_0, A_1 in X. Do the same to A_0 to obtain A_{00} and A_{01} and to A_1 to get A_{10} and A_{11}. Repeat indefinitely. Let $F = (A_0 \cup A_1) \cap (A_{00} \cup A_{01} \cup A_{10} \cup A_{11}) \cap \cdots$. Let $f : F \to \mathbb{R}$ be defined by taking x to $.b_1 b_2 b_3 \ldots$ where the b_i are binary digits, provided $x \in A_{b_1} \cap A_{b_1 b_2} \cap A_{b_1 b_2 b_3} \cap \cdots$.)

9. Show that $\{x_n\} \subseteq \mathbb{C}^\wedge$ converges to x_0 in the ρ_c metric if and only if either $x_0 \neq \infty$ and $|x_n - x_0| \to 0$ or $x_0 = \infty$ and $\frac{1}{|x_n|} \to 0$.

10. Prove the remark following (1.22):

$$|\rho(x,y) - \rho(x',y')| \leq \rho(x,x') + \rho(y,y')$$

and deduce the continuity of ρ.

11. Prove Proposition (1.26). Then show that a continuous real-valued function on a compact space attains its maximum. (Use the fact that a bounded closed subset of $\mathbb{R}$ contains its supremum.)

§2. Preliminaries: Vector Spaces and Complex Variables

Our goal is the treatment of complex analysis from the point of view of topological vector spaces. Here we present the basic definitions and some well-known facts.

(2.1) Definition. A vector space over $\mathbb{C}$ is a set E together with the operations of addition $(+)$ and scalar multiplication satisfying

(a) E is an Abelian group under addition. That is, for any $a, b \in E$ the equation $a + x = b$ has a unique solution in E. Moreover $a + b = b + a$ and $a + (b + c) = (a + b) + c$ for all $a, b, c \in E$. The solution of $a + x = a$ is denoted 0. For $a + x = 0$ we write $x = -a$.

(b) For $\alpha \in \mathbb{C}$ and $x \in E$, the product $\alpha x \in E$ is defined and satisfies for all $\alpha, \beta \in \mathbb{C}$, $x, y \in E$:

$$\alpha(x + y) = \alpha x + \alpha y, \ (\alpha + \beta)x = \alpha x + \beta x, \ \alpha(\beta x) = (\alpha\beta)x,$$

and $1x = x$.

The same definition could be made for any field in place of $\mathbb{C}$, for example $\mathbb{R}$.

(2.2) Definition. A vector space E over $\mathbb{C}$ is called an algebra over $\mathbb{C}$ if it has a multiplication $(x,y) \to xy$ which satisfies.

(c) For all $x, y, z \in E$ and $\alpha \in \mathbb{C}$, $(xy)z = x(yz)$, $x(y+z) = xy = xz$, $(x + y)z = xz + yz$ and $\alpha(xy) = (\alpha x)y = x(\alpha y)$.

(2.3) Definition. (E,Ω) is a topological vector space if E is a vector space and Ω is a Hausdorff topology on E such that addition and scalar multiplication are continuous (addition from $E \times E$ to E and scalar multiplication from $\mathbb{C} \times E$ to E. The product topology is assumed on $E \times E$ and $\mathbb{C} \times E$). If E is an algebra and multiplication is also continuous, (E,Ω) is called a topological algebra.

We usually just write E instead of (E,Ω).

(2.4) Definition. If E is a vector space and $p : E \to [0,\infty)$, then *p is called a seminorm if* $p(\alpha x) = |\alpha| p(x)$ *for each* $\alpha \in \mathbb{C}$, $x \in E$, *and* $p(x + y) \leq p(x) + p(y)$ *for all* $x, y \in E$. *If in addition* $p(x) > 0$ *when* $x \neq 0$, *then* p *is called a norm.*

(2.5) *Definition. A topological vector space* E *is said to be locally convex if there exists a family of seminorms* P *such that the sets* $\{x \in E : p(x - y) < \varepsilon\}$ $p \in P$, $y \in E$, $\varepsilon > 0\}$ *generate the topology on* E.

The reason for the term *locally convex* is that the sets $B(y,p,\varepsilon) = \{x \in E : p(x - y) < \varepsilon\}$ are *convex*. That is, if x, $x' \in B(y,p,\varepsilon)$ and $0 < \lambda < 1$, then $\lambda x + (1 - \lambda)x' \in B(y,p,\varepsilon)$

(2.6) *Important Example. If* G *is an open subset of* $\mathbb{C}$ *(or* $\mathbb{C}^\wedge$*) let* $C(G)$ *be the vector space of continuous complex-valued functions on* G. *Let* $P = \{p_K\}$ *where* K *runs over compact subsets of* G, *writing* $p_K(f) = \|f\|_K = \sup\{|f(z)| : z \in K\}$.

(2.7) *Our Hero.* $H(G)$ *is the subspace of* $C(G)$ *consisting of all functions holomorphic on* G, *with the same definition of the seminorms* $\|f\|_K$.

Our definition of holomorphic is the existence of the limit $\lim_{z \to a} \frac{f(z) - f(a)}{z - a} = f'(a)$ for all $a \in G$. It is clear that $H(G)$ is a subspace of $C(G)$ (i.e. it is closed under addition and scalar multiplication) and that it is a subalgebra (it is also closed under multiplication). A substantial part of our work will be the study of $H(G)$. The rest of this chapter will be devoted to a review of some standard results about holomorphic functions, with one or two less standard results about continuous functions.

(2.8) *Invariance of Angle. Let* f *be holomorphic in a neighborhood of* $z_0 \in \mathbb{C}$, *let* α *and* β *be two rays emanating from* z_0, *and let* α', β' *be the images under* f *of* α, β. *If* $f'(z_0) \neq 0$, *then the angle between* α' *and* β' *at* $f(z_0)$ *equals the angle between* α *and* β.

Proof. Let $z_0 + at$ $(z_0 + bt)$ be the formula for α (resp. β). Then the equation of α' is $g(t) = f(z_0 + at)$ and the tangent line to α' at $f(z_0)$ has the formula $f(z_0) + g'(0)t$. Since $g'(0) = f'(z_0)a$, we see that the direction of α is rotated by an angle

$\arg f'(z_0)$. The direction of β will be rotated by the same angle. QED

(2.9) Definition. A curve $\gamma : [0,1] \to \mathbb{C}$ is said to be rectifiable if $\sup \sum_{i=1}^{n} |\gamma(t_i) - \gamma(t_{i-1})| < +\infty$. The supremum is taken over all partitions $0 = t_0 < t_1 < \cdots < t_n = 1$ of the unit interval. The value of the supremum is called the length of the curve and is denoted $\|\gamma\|$.

If γ is a rectifiable curve and f is a continuous function on the image $\gamma\hat{}$, then $\int_\gamma f(z)dz = \lim \sum f(\gamma(\xi_i))(\gamma(t_i) - \gamma(t_{i-1}))$. The meaning of this limit is the following. There exists a unique number $I(f,\gamma)$ such that for every $\varepsilon > 0$ there exists a $\delta > 0$ such that

$$|I(f,\gamma) - \sum_{i=1}^{n} f(\gamma(\xi_i))(\gamma(t_i) - \gamma(t_{i-1}))| < \varepsilon$$

whenever $\{0 = t_0 < t_1 < \cdots < t_n = 1\}$ is a partition of $[0,1]$ with $\max_i |t_i - t_{i-1}| < \delta$ and $\xi_i \in [t_{i-1}, t_i]$. If γ has a continuous derivative $\gamma'(t)$, then $\int_\gamma f(z)dz = \int_0^1 f(\gamma(t))\gamma'(t)dt$. If γ_1 and γ_2 are equivalent curves (i.e. $\gamma_2(t) = \gamma_1(\lambda(t))$ where $\lambda : [0,1] \to [0,1]$ is onto and strictly increasing) then $\int_{\gamma_1} f(z)dz = \int_{\gamma_2} f(z)dz$. We will need this only in the case where γ_1 and γ_2 have continuous, nowhere vanishing derivatives. In this case λ is continuously differentiable and then

$$\int_{\gamma_2} f(z)dz = \int_0^1 f(\gamma_1(\lambda(t)))\gamma_1'(\lambda(t))\lambda'(t)dt$$
$$= \int_0^1 f(\gamma_1(t))\gamma_1'(t)dt = \int_{\gamma_1} f(z)dz.$$

If F is holomorphic on a neighborhood of $\gamma\hat{}$, then $\int_\gamma F'(z)dz = F(\gamma(1)) - F(\gamma(0))$. In particular if f is holomorphic in a domain U (an open connected set in $\mathbb{C}$) and has a primitive F (i.e. $F' = f$) and if γ is a simple closed curve in U then

$$\int_\gamma f(z)dz = \int_\gamma F'(z)dz = F(\gamma(1)) - F(\gamma(0)) = 0.$$

If follows from the definition of integration that $|\int_\gamma f(z)dz| \leq \|\gamma\| \cdot \max_\gamma |f(z)|$. If γ_1 and γ_2 are curves with $\gamma_1(1) = \gamma_2(0)$, then $\gamma_1 + \gamma_2$ is sometimes used to denote the path obtained by laying γ_1 and γ_2 end to end. That is $\gamma = \gamma_1 + \gamma_2$ is defined by $\gamma(t) = \gamma_1(2t)$, $0 \le t \le \frac{1}{2}$, and $\gamma(t) = \gamma_2(2t - 1)$, $\frac{1}{2} \le t \le 1$. Then $\int_{\gamma_1+\gamma_2} f(z)dz = \int_{\gamma_1} f(z)dz + \int_{\gamma_2} f(z)dz$.

(2.10) Cauchy's Theorem for Rectangles. Let f be holomorphic in a neighborhood of the rectangle $R = [a,b] \times [c,d]$ and let ∂R denote the boundary of this rectangle; then $\int_{\partial R} f(z)dz = 0$.

Proof. Bisect the sides of the rectangle to form four rectangles R^1, R^2, R^3 and R^4. Without loss of generality we may assume ∂R is parametrized by $\gamma(t)$ so that ∂R is traversed counter-clockwise as t runs from 0 to 1. With the same convention on the ∂R^j it follows that $\int_{\partial R} f(z)dz = \sum_{j=1}^{4} \int_{\partial R^j} f(z)dz$. Thus there exists an R^j, denoted R_1, such that $|\int_{\partial R} f(z)dz| \le 4|\int_{\partial R_1} f(z)dz|$. Dissecting R_1 in the same way leads to

$$\left|\int_{\partial R} f(z)dz\right| \le 4^2 \left|\int_{\partial R_2} f(z)dz\right|$$

and, by induction

$$\left|\int_{\partial R} f(z)dz\right| \le 4^n \left|\int_{\partial R_n} f(z)dz\right|$$

where $\{R_n\}$ is a sequence of rectangles with $R_{n+1} \subset R_n$ and $\|\partial R_n\| = 2\|\partial R_{n+1}\|$. Let z_0 be the unique element of $\cap R_n$. By the definition of f' it follows that $f(z) = f(z_0) + f'(z_0)(z - z_0) + (z - z_0)h(z)$, where $h(z) \to 0$ as $z \to z_0$. Both constant functions and the identity function z have primitives so $\int_{\partial R_n} f(z_0)dz = \int_{\partial R_n} f'(z_0)(z - z_0)dz = 0$. Thus $|\int_{\partial R_n} f(z)dz| = |\int_{\partial R_n} (z - z_0)h(z)dz| \le \|\partial R_n\| \cdot \delta(R_n)\max_{R_n}|h(z)| = 4^{-n}\|R\|\delta(R)\max_{R_n}|h(z)|$ and so $|\int_{\partial R} f(z)dz| \le \|R\|\delta(R)\max_{R_n}|h(z)| \to 0$, since $h(z) \to 0$ as $z \to z_0$. QED

Cauchy's Theorem for the rectangle can be used to prove the stronger result: $\int_{\partial R} f(z)dz = 0$ if f is holomorphic in a neighborhood of R except at one point where it is continuous. Simply enclose that one point in a small rectangle R' and dissect $R \setminus R'$ into rectangles. Then $\int_{\partial R'} f(z)dz$ can be made arbitrarily small and $\int_{\partial(R\setminus R')} f(z)dz = 0$ by (2.10). Thus, defining $g(z) = \frac{f(z) - f(a)}{z - a}$, $z \neq a$; $g(a) = f'(a)$ we see that $\int_R g(z)dz = 0$ if f is holomorphic in a neighborhood of R. This leads to

(2.11) Cauchy Integral Formula for a Rectangle. *If* R *is a rectangle and* $a \in \text{int } R$, *then*

$$\frac{1}{2\pi i}\int_{\partial R} \frac{f(z)}{z - a}\, dz = f(a).$$

Proof. By the above, $\frac{1}{2\pi i}\int_{\partial R} \frac{f(z)}{z - a}\, dz = \frac{f(a)}{2\pi i}\int_{\partial R} \frac{dz}{z - a}$. It remains only to prove that $\int_{\partial R}(z - a)^{-1}dz = 2\pi i$. Without loss of generality take $a = 0$. Then

$$\int_{\partial R} \frac{1}{z}\, dz = \int_{\partial R} \frac{x - iy}{x^2 + y^2}(dx + idy)$$

$$= \int_{\partial R} \frac{xdx + ydy}{x^2 + y^2} + i\int_{\partial R} \frac{xdy - ydx}{x^2 + y^2}.$$

Apply Green's Theorem to the domain $\text{int } R\setminus D^-$ where D^- is a closed disk centered at 0. This gives

$$\int_{\partial R} \frac{1}{z}\, dz = \int_{\partial D} \frac{1}{z}\, dz.$$

Parametrize ∂D by $\gamma(t) = re^{2\pi it}$ where r is the radius of D. Then $\int_{\partial D} z^{-1}dz = \int_0^1 2\pi i\, dt = 2\pi i$. QED

(2.12) Proposition. *If* f *is holomorphic in a disk* D, *then* f *has a primitive in* D.

Proof. Define $F(w) = \int_{\gamma_w} f(z)dz$ where γ_w is any curve from the center of D to w which consists entirely of horizontal and vertical line segments. From (2.10) $F(w)$ is independent of the choice of curve. Then $\frac{F(w) - F(w')}{w - w'} = \frac{1}{w - w'}\int_{w'}^{w} f(z)dz$ where the integral is along the the curve from w' to w consisting of one horizontal, then one vertical line segment. Let $w' \to w$ along first a horizontal and then a vertical line segment to verify first the Cauchy-Riemann equations for F and then that $F' = f$. QED

Thus, if f is holomorphic in a disk D and γ is any rectifiable simple closed curve, $\int_\gamma f(z)dz = 0$. Similarly

$$\frac{1}{2\pi i}\int_\gamma \frac{f(z)}{z-a}\,dz = f(a)$$

if γ is any simple closed curve in D and a is any point with $\int_\gamma (z-a)^{-1}dz = 2\pi i$. Using Green's Theorem as in (2.11) the reader can show that if A is an open set in D with $\gamma^\wedge = \partial A \subset D$ and γ is a simple closed curve with continuous derivative, then $f(a) = \frac{1}{2\pi i}\int_\gamma f(z)(z-a)^{-1}dz$ whenever $a \in A$. This is all we need to prove the following.

(2.13) Power Series Representation. *If f is holomorphic in the unit disk $\mathbb{D} = \{z : |z| < 1\}$ then $f(z) = \sum_{n=0}^{\infty} a_n z^n$ where $2\pi i a_n = \int_\gamma f(z)z^{-n-1}dz$ for $\gamma(t) = re^{2\pi i t}$, $0 \le t \le 1$, and any $r \in (0,1)$. Moreover the series converges absolutely for all $z \in \mathbb{D}$ and uniformly for $|z| < r$ when $r < 1$.*

Proof. Write $\frac{1}{\zeta - z} = \frac{1}{\zeta}\sum_{n=0}^{\infty}\left(\frac{z}{\zeta}\right)^n$. If $|z| = \rho$ then this series converges uniformly for ζ on $\{\zeta : |\zeta| = r\}$ when $\rho < r < 1$. Using the Cauchy Integral Formula as extended above we get

$$\begin{aligned} f(z) &= \frac{1}{2\pi i}\int_{|\zeta|=r}\frac{f(\zeta)}{\zeta - z}\,d\zeta \\ &= \sum_{n=0}^{\infty} z^n \frac{1}{2\pi i}\int_{|\zeta|=r}\frac{f(\zeta)}{\zeta^{n+1}}\,d\zeta \\ &= \sum_{n=0}^{\infty} a_n z^n . \end{aligned}$$

Since power series are differentiable to any order and the coefficients are uniquely determined, it follows that the a_n are independent of the choice of r and $n!a_n = f^{(n)}(0)$. Since the power series converges in $|z| < r$ for arbitrary $r < 1$, the rest of the statements are standard facts about power series. QED

It follows from the proof of (2.13) that if f is holomorphic in a neighborhood of z_0 then $f(z) = \sum_{n=0}^{\infty} a_n(z-z_0)^n$ where $a_n = \frac{1}{n!}f^{(n)}(z_0)$ and the power series converges in the largest disk about z_0 to which f can be extended as a holomorphic function. The formulae for the coefficients in (2.13) give the following.

(2.14) Cauchy Inequalities. *If $\sum_{n=0}^{\infty} a_n z^n = f(z)$ converges in $|z| < R$,* then for $r < R$,

$$|a_n| \leq \frac{1}{r^n} \max_{|z|=r} |f(z)|.$$

(2.15) Uniqueness Theorem. *If f is holomorphic in a connected open set U and $f(z_n) = 0$ for a sequence $\{z_n\}$ in U with $z_n \to z_0 \in U$, then $f(z) \equiv 0$ in U.*

Proof: Writing $f(z) = \sum_{n=0}^{\infty} a_n(z - z_0)^n$ in a neighborhood of z_0 it follows that $a_0 = 0$. Then $f(z)/(z - z_0) = a_1 + a_2(z - z_0) + \cdots$ and it follows that $a_1 = 0$. Continuing in this manner we see that $a_n = 0$ for all n and so $f(z) \equiv 0$ in a neighborhood of z_0, i.e. we have shown that the set of limit points of $f^{-1}(0)$ is an open set in U. Since it is clearly also closed and U is connected, it must be all of U. QED

(2.16) Corollary. *The zeros of f in U are isolated*, i.e. *If f is not identically zero and $f(z_0) = 0$ then there is a neighborhood N of z_0 such that $f(z) \neq 0$ in N except at $z = z_0$.*

If f is holomorphic in an open set U containing z_0 and $f(z_0) = 0$, then f is said to have a zero of order m if $f(z) = a_m(z - z_0)^m + a_{m+1}(z - z_0)^{m+1} + \cdots$, where $a_m \neq 0$. That is $f^{(m)}(z_0) \neq 0$ but $f^{(n)}(z_0) = 0$ for $n < m$.

(2.17) Proposition. *Let R be an open rectangle and let f be analytic in an open set containing $\bar{R}$. Assume that f is never zero on ∂R. Then $(1/2\pi i)\int_{\partial R} f'(z)/f(z)dz = N$, where N is the number of zeros of f in R, each counted as many times as its order.*

Proof. Since f is not identically zero it has finitely many zeros in R. Denote them by $z_1, z_2, \ldots, z_m$. Let $R_1, R_2, \ldots, R_m$ be open rectangles (with sides parallel to those of R) such that $z_j \in R_j$, $\bar{R}_i \cap \bar{R}_j = \emptyset$ when $i \neq j$ and $\bar{R}_j \subset R$ for all j, By extending the sides of the R_j we can decompose R into rectangles $S_1, \ldots, S_n$ among which are included the R_j. Clearly

$$\begin{aligned}\int_{\partial R} f'(z)/f(z)dz &= \sum_k \int_{\partial S_k} f'(z)/f(z)dz \\ &= \sum_j \int_{\partial R_j} f'(z)/f(z)dz\end{aligned}$$

since $\int_{\partial S_k} f'(z)/f(z)dz = 0$ if f has no zero in S_k^-. It now suffices to show that $(1/2\pi i)\int_{\partial R_j} f'(z)/f(z)dz$ is the order of the zero z_j. Now in R_j^-, $f(z)$ can be written in the form $f(z) = (z - z_j)^{m_j} g(z)$ where g has no zeros in R_j^- and m_j is the order of z_j. Then $f'(z)/f(z) = m_j(z - z_j)^{-1} + g'(z)/g(z)$. The integral of the first term over ∂R_j is $2\pi i m_j$ and that of the second is zero. QED

(2.18) Open Mapping Theorem. Let U be an open subset of $\mathbb{C}$ and f a function which is holomorphic in U and non-constant in any component of U; then $f(U)$ is open.

Proof. We remark that the components of U are open sets so we may suppose U is connected. Let $w_0 = f(z_0) \in f(U)$. Since f is not constant $f(z) - w_0$ has an isolated zero at z_0. Let R be an open rectangle containing z_0 with $R^- \subset U$, such that $f(z) \neq w_0$ in R^- except at z_0. Then the function $N(w) = \frac{1}{2\pi i}\int_{\partial R} f'(z)/(f(z)-w)dz$ is clearly continuous for $w \in R$ and, by (2.17), non-zero at $w = w_0$. Thus $N(w)$ is non-zero in a neighborhood B of w_0. But $N(w)$ is the number of solutions in R of $f(z) = w$, so every $w \in B$ satisfies $f(z) = w$ for some $z \in R$, i.e. $B \subset f(U)$. Thus $f(U)$ is open. QED

We note that the proof actually shows a little more: If $f(z) - w_0$ has a zero of order N at z_0, then there is a neighborhood V of w_0 and a rectangle R containing z_0 such that for each $w \in V$, $f(z) - w$ has N zeros in R. The sets R and V can be chosen so that, except when $w = w_0$, $f(z) = w$ has N distinct zeros in R. Indeed, choose R so small that $f'(z)$ is not zero in R^- except at z_0. (This can be done because the zeros of f' must have no limit point in U.) Then if $f(z) - w = 0$ and $w \neq w_0$ we must have $\frac{d}{dz}(f(z) - w) = f'(z) \neq 0$. Thus the zeros of $f(z) - w$ are simple (i.e. of order 1). Let us state this as a Proposition:

(2.19) Proposition. Let U be an open set in $\mathbb{C}$ and $f \in H(U)$. Let $z_0 \in U$ with $f(z_0) = w_0$ and suppose $f(z) - w_0$ has a zero at z_0 of order N. Then there is a neighborhood R of z_0 and a neighborhood V of w_0 such that for each $w \in V$, $w \neq w_0$, $f(z) - w$ has exactly N distinct solutions in R.

(2.20) Corollary. If $f \in H(U)$ is one-to-one, then $f'(z) \neq 0$ for any $z \in U$.

Proof: If $f'(z_0) = 0$ then $f(z) - f(z_0)$ has a zero of order at least 2. Thus, for some w near $f(z_0)$, $f(z) = w$ has at least two solutions in U. This contradicts the fact that f is one-to-one. QED

(2.21) Maximum Modulus Principle. If U is an open connected set and f is holomorphic in U, then $|f|$ does not have a maximum in U unless f is constant.

Proof. If f is not constant then $f(U)$ is open. If $z_0 \in U$ then $f(U)$ contains a disk around $f(z_0)$ and so there exist points $z \in U$ with $|f(z)| > |f(z_0)|$. Thus no $z_0 \in U$ can be a maximum for $|f|$. QED

(2.22) Definitions. (Isolated singularity) If f is holomorphic in a neighborhood of z_0 except at z_0 itself, z_0 is called an isolated singularity of f.

(Removable singularity) If $f(z_0)$ may be defined so that f is holomorphic in a neighborhood of z_0, then z_0 is said to be a removable singularity.

(Pole) If $\lim_{z\to z_0}|f(z)| = \infty$ then z_0 is called a pole of f.

(Essential Singularity) If z_0 is an isolated singularity but neither a pole nor a removable singularity, it is called an essential singularity.

(2.23) Laurent Series Representation. Suppose f is holomorphic in the deleted disk $\{z : 0 < |z| < 1\}$. Then $f(z) = \sum_{n=-\infty}^{\infty} a_n z^n$ where $a_n = \frac{1}{2\pi i}\int_{|z|=r} f(z)z^{-n-1}dz$, $n = 0, \pm 1, \ldots,$ and $0 < r < 1$. The series converges uniformly in any annulus $\{z : r \leq |z| \leq R\}$ with $0 < r < R < 1$.

Proof. Let z satisfy $0 < |z| < 1$ and let $0 < r < |z| < R < 1$. Let $\gamma(t) = re^{2\pi it}$, $0 \leq t \leq 1$, and $\Gamma(t) = Re^{2\pi it}$, $0 \leq t \leq 1$. We claim

$$(1) \qquad f(z) = \frac{1}{2\pi i}\int_{\Gamma} \frac{f(\zeta)}{\zeta - z}\,d\zeta - \frac{1}{2\pi i}\int_{\gamma} \frac{f(\zeta)}{\zeta - z}\,d\zeta.$$

This follows from the remarks after (2.12) via the following argument. By drawing in segments of radii between $\gamma^\wedge$ and $\Gamma^\wedge$ and additional circles with radii between r and R we can write

$$\frac{1}{2\pi i}\int_\Gamma \frac{f(\zeta)}{\zeta - z}\,d\zeta - \frac{1}{2\pi i}\int_\gamma \frac{f(\zeta)}{\zeta - z}\,d\zeta = \sum_j \frac{1}{2\pi i}\int_{\partial A_j} \frac{f(\zeta)}{\zeta - z}\,d\zeta,$$

where each A_j is contained in a disk contained in $0 < |\zeta| < 1$. Every term in the sum is zero except perhaps when $z \in A_j$. That term equals $f(z)$.

Write $\frac{1}{\zeta - z} = \sum_{n=0}^{\infty} z^n \zeta^{-n-1}$ when $\zeta \in \Gamma^\wedge$ and

$$\frac{1}{\zeta - z} = -\sum_{n=0}^{\infty} \zeta^n z^{-n-1} = -\sum_{n=-\infty}^{-1} z^n \zeta^{-n-1} \quad \text{when} \quad \zeta \in \gamma^\wedge.$$

Insert these in (1) to get

(2) $$f(z) = \sum_{n=0}^{\infty} z^n \frac{1}{2\pi i}\int_\Gamma f(\zeta)\zeta^{-n-1}d\zeta + \sum_{n=-\infty}^{-1} z^n \frac{1}{2\pi i}\int_\gamma f(\zeta)\zeta^{-n-1}d\zeta.$$

Both series for $(\zeta - z)^{-1}$ converge uniformly in z for $r_1 \le |z| \le R_1$ when $r < r_1$ and $R_1 < R_1$ and so the series (2) converges in $r_1 \le |z| \le R_1$. The expression $\frac{1}{2\pi i}\int_\gamma f(\zeta)\zeta^{-n-1}d\zeta$ is independent of the choice of r by an argument similar to the one establishing (1). QED

(2.24) Proposition. *If f has an isolated singularity at z_0 then $f(z) = \sum_{-\infty}^{\infty} a_n(z - z_0)^n$ in some deleted neighborhood of z_0. f has a removable singularity at z_0 if and only if $a_n = 0$ for all $n < 0$. f has a pole at z_0 if and only if for some $m < 0$, $a_m \ne 0$ and $a_n = 0$ for all $n < m$. f has an essential singularity if and only if $a_n \ne 0$ for infinitely many $n < 0$.*

Proof. The series representation follows from (2.23). If $a_n = 0$ for all $n < 0$, then defining $f(z_0) = a_0$ will make f holomorphic in a neighborhood of z_0. Conversely if f is holomorphic in a neighborhood of z_0 then f has a power series representation, i.e. $a_n = 0$ for $n < 0$.

If f is bounded near z_0 then

$$2\pi i\, a_n = \lim_{r\to 0} \int_{|z-z_0|=r} f(z)z^{-n-1}dz = 0 \quad \text{for} \quad n < 0.$$

If f has a pole at z_0 then this shows that $1/f$ has a removable singularity at z_0. Let m be the order of the zero of $1/f$ at z_0. Then $1/f(z) = (z - z_0)^m g(z)$ where g is holomorphic in a neighborhood of z_0 and $g(z_0) \ne 0$. Write $1/g(z) = b_0 + b_1(z - z_0) + \cdots$. Then $f(z) = b_0(z - z_0)^{-m} + b_1(z - z_0)^{-m+1} + \cdots$. Conversely, it is clear that if $f(z) = a_{-m}(z - z_0)^{-m} + a_{-m+1}(z - z_0)^{-m+1} + \cdots$, then $\lim_{z \to z_0} |f(z)| = \infty$.

Finally, it is clear by the process of elimination that f has an essential singularity if and only if $a_n \neq 0$ for infinitely many $n < 0$. QED

(2.25) Casorati-Weierstrass Theorem. If f is holomorphic in $\{z : 0 < |z - z_0| < r\}$ and has an essential singularity at z_0 then $f(0 < |z - z_0| < \varepsilon)$ is dense in $\mathbb{C}$ for every $\varepsilon > 0$.

Proof. If not, then for some $a \in \mathbb{C}$, $1/(f(z) - a)$ is bounded near z_0. Consequently it has a removable singularity at z_0. This implies that f has a removable singularity or pole at z_0. QED

(2.26) Definition. If f has a pole at z_0, $f(z) = a_{-m}(z - z_0)^{-m} + \cdots$ and $a_{-m} \neq 0$, then f is said to have a pole at z_0 of order m.

(2.27) Proposition. f has a pole at z_0 of order m if and only if $1/f$ has a zero at z_0 of order m.

(2.28) Liouville's Theorem. If f is holomorphic in $\mathbb{C}$ and f is bounded, then f is constant.

Proof. $f(z) = \sum_{n=0}^{\infty} a_n z^n$ for all $z \in \mathbb{C}$.

Moreover $f(\frac{1}{z})$ has a removable singularity at 0 by the proof of (2.24), so $a_n = 0$ for $n > 0$. Thus $f(z) \equiv a_0$. QED

We end the preliminary chapters with two famous theorems about continuous functions which we state without proof. The reader is referred to [Kelley].

(2.29) Stone-Weierstrass Theorem. Let X be a compact Hausdorff space and let $C(X)$ be the space of continuous complex valued functions on X with the norm $\|f\|_X = \sup\{|f(x)| : x \in X\}$. Let A be a subalgebra of $C(X)$ such that $f \in A$ implies $\bar{f} \in A$ and A contains all constant functions. Then A is dense in $C(X)$ if A separates points on X (that is, for any $x, y \in X$, $x \neq y$, there exists $f \in A$ with $f(x) \neq f(y)$).

(2.30) Tietze's Extension Theorem. Let X be a normal topological space, that is, for any two disjoint closed sets E and F in X there exist disjoint open sets U, V with $E \subset U$, $F \subset V$. Then for any closed set $A \subset X$ and any continuous function $f : A \to \mathbb{C}$, there exists a continuous $F : X \to \mathbb{C}$ such that $F(x) = f(x)$ for $x \in A$. Moreover, if $|f(x)| \leq 1$ for all $x \in A$ then F can be chosen with $|F(x)| \leq 1$ for all $x \in X$.

Examples of normal spaces are metric spaces and compact Hausdorff spaces.

NOTES: Two references for further study of topological vector spaces include [J. L. Kelley and I. Namioka] and [G. Köthe, 2].

For complex analysis see [C. Carathéodory, 1], [E. Hille], and [S. Saks and A. Zygmund].

Exercises

1. Show that a nearly identical proof to (2.10) would prove a similar theorem for triangles, and that all the consequences of (2.10) are consequences of such a theorem. What difficulties (if any) occur in trying to prove Cauchy's Theorem for circles directly as in (2.10)?

2. Try to prove $\int_{\partial R} \frac{1}{z}\, dz = 2\pi i$, where R is a rectangle containing 0, by direct integration, thus avoiding Green's Theorem in (2.11).

3. Let E be the vector space consisting of sequences $\{a_n\}_0^\infty$ of complex numbers such that $\sum |a_n| r^n$ converges for every $r > 0$. Define two collections of seminorms on E. $P = \{p_r : r > 0\}$, where $p_r(\{a_n\}) = \sum |a_n| r^n$, and $P' = \{p'_r : r > 0\}$, where $p'_r(\{a_n\}) = \sup_n |a_n| r^n$. Show that a sequence $\alpha_k = \{a_n^{(k)}\}_{n=1}^\infty$, $k = 1,2,\ldots$ converges to 0 with respect to P if and only if it converges to 0 with respect to P', i.e. $p_r(\alpha_k) \to 0$ for all r if and only if $p'_r(\alpha_k) \to 0$ for all r.

4. Let $T : E \to H(\mathbb{C})$ (E as in problem 3) be defined by $T(\{a_n\}) = \sum a_n z^n$. Show that T is one-to-one and onto. Show that if $\{\alpha_k\} \subseteq E$ is a sequence then $p_r(\alpha_k) \to 0$ for each r if and only if $T(\alpha_k) \to 0$ uniformly on each compact set in $\mathbb{C}$.

5. If $\mathbb{D} = \{z : |z| < 1\}$ and if $H(\mathbb{D})$ is topologized by the seminorms $\|\cdot\|_K$ as in (2.7), show that the function $a_n(f)$ taking a function $f \in H(\mathbb{D})$ to the nth coefficient a_n of the Maclaurin series

of f (i.e. The function $\sum_k a_k z^k \mapsto a_n$) is continuous. (By linearity it is enough to show that if $\|f_m\|_K \to 0$ for each compact $K \subseteq D$, then $a_n(f_m) \to 0$ as $m \to \infty$. Use (2.14).)

6. If f is holomorphic in $|z| > R$, use (2.24) to show that $f(z) = \sum_{n=-\infty}^{\infty} a_n z^n$ with convergence in $|z| > R$. f is said to be <u>analytic at</u> ∞ if $a_n = 0$ for all $n > 0$, then $f(\infty)$ is set equal to a_0 and $f'(\infty) = a_{-1}$. Show that f is analytic at ∞ iff $g(z) = f(\frac{1}{z})$ has a removable singularity at 0, that $f(\infty) = g(0)$ and $f'(\infty) = g'(0) = \lim_{|z|\to\infty} z(f(z) - f(\infty))$. Formulate the proper definition of "∞ is an essential singularity of f" and prove the analogue of (2.25).

7. Prove the following generalization of Liouville's Theorem (2.28): If f is holomorphic in $\mathbb{C}$ and if for some $m \geq 0$ and $C > 0$, $|f(z)| \leq C|z|^m$ for all $z \in \mathbb{C}$, $|z| \geq 1$, then f is a polynomial of degree at most m.

8. Show that Tietze's Extension Theorem is false if "continuous" is replaced by "holomorphic". That is, show that in general a function which is holomorphic on some closed set $A \subseteq \mathbb{C}$ (i.e. holomorphic in a neighborhood of A) need not have an extension which is holomorphic in all of $\mathbb{C}$.

9. Let X be a closed line segment in $\mathbb{C}$ and let A be the set of restrictions to X of functions holomorphic in $\mathbb{C}$. Show that A is dense in $C(X)$.

10. Let K, L be two closed sets in $\mathbb{C}$ with $K \cap L = \emptyset$. Find an explicit example of a continuous function f on $\mathbb{C}$ such that $f(z) = 1$ if $z \in K$ and $f(z) = -1$ if $z \in L$ and $-1 \leq f(z) \leq 1$ for all $z \in \mathbb{C}$. (The existence of f is guaranteed by (2.30).) Hint: Make use of the functions $\rho(z,K)$ and $\rho(z,L)$ where ρ denotes distance.

11. Show that the inverse of a one-to-one holomorphic function is holomorphic. Hint: If f is holomorphic in G and $g : f(G) \to G$ is its inverse then g is continuous (why?) and $\lim_{w\to b} \frac{g(w) - g(b)}{w - b} = \lim_{z\to g(b)} \frac{z - g(b)}{f(z) - f(g(b))}$. The latter limit exists because $f'(a) \neq 0$ for any a (why?).

12. Let $f \in H(G)$ where G is an open set in $\mathbb{C}$. Suppose for every $z \in G$ there is a positive integer $n = n_z$ such that $f^{(n)}(z) = f(z)$. Show that f has an extension to $H(\mathbb{C})$. Hint: Baire Category and then the Uniqueness Theorem (or do it without Baire Category.)

13. Let $\varphi : [0,+\infty) \to [0,+\infty)$ be any continuous function. Prove that there is a function f in $H(\mathbb{C})$ such that $f(x) \geq \varphi(x)$ for all $x \in [0,+\infty)$. (Hint: $f(z) = \sum_{n=0}^{\infty} a_n z^n$ with properly chosen a_n.)

14. Let $f \in H(\mathbb{C})$ be such that $\mathbb{C} \setminus f(\mathbb{C})$ contains a ray. Prove that f is a constant. (Use Liouville's Theorem.)

15. a) Show that $f \in H(G)$ has a primitive if and only if $\int_\gamma f(z)dz = 0$ for every closed curve γ in G.

b) Suppose f has no zeros in G and that h is a primitive of f'/f. Show that $e^h = \text{const } f$ (Hint: Compute $(fe^{-h})'$).

c) Show that there is an analytic function $h \in H(G)$ such that $\exp h(z) = z$ if and only if $\int_\gamma z^{-1}dz = 0$ for every closed curve γ in G.

16. Suppose $f \in H(\mathbb{C})$ and suppose for every positive integer n there is a $g_n \in H(\mathbb{C})$ such that $g_n^n = f$. Show that either $f \equiv 0$ or $f = \exp g$ for some $g \in H(\mathbb{C})$. (Hint: $\int_\gamma f'/f = n \int_\gamma g_n'/g_n$. Then use (2.17) and exercise 15 a,b.)

17. Prove that a one-to-one entire function $f(z)$ has the form $f(z) = az + b$. (Hint: Use (2.25) to show that f cannot have an essential singularity at ∞ and so f is a polynomial.)

§3. Properties of $C(G)$ and $H(G)$

With little change we can study functions of several variables. To keep the notation simple, we will restrict ourselves to two variables. In that case, G is an open set in $\mathbb{C} \times \mathbb{C} = \mathbb{C}^2$. Then $C(G)$ is defined just as in the one-variable case. (The distance in $\mathbb{C}^2$ between two points (z,w) and (z',w') will be denoted $d((z,w), (z',w')) = (|z - z'|^2 + |w - w'|^2)^{\frac{1}{2}}$). We also define $H(G)$ as before, but must first define holomorphic.

(3.1) Definition: f is holomorphic at a point $p_0 = (z_0, w_0)$ provided that f is continuous in some neighborhood N of p_0 and is holomorphic in each variable separately in N.

This last statement means the following: For each fixed $w \in \mathbb{C}$ the function $g_w(z) = f(z,w)$ is holomorphic at all z for which $(z,w) \in N$. And similarly $h_z(w) = f(z,w)$ is holomorphic in w for each fixed z. The symbols $\partial f/\partial z$ and $\partial f/\partial w$ will stand for g_w' and h_z' respectively, in analogy with the real variable situation. A theorem of Hartogs shows that if $\partial f/\partial z$ and $\partial f/\partial w$ both exist at each point in G then f is holomorphic in G, i.e. the continuity assumption is superfluous. We will not prove this here.

Our first goal is to prove that $C(G)$ is metrizable. For this we need a definition.

(3.2) Definition. A sequence $K_1, K_2, K_3, \ldots$ of compact subsets of G is exhaustive if for each compact set $K \subseteq G$, there is some index n such that $K \subseteq K_n$.

It is an easy exercise to show that sufficient conditions for $\{K_n\}$ to be exhaustive are (i) $\bigcup_n K_n = G$ and (ii) $K_n \subseteq \operatorname{int} K_{n+1}$. Thus for any open set G, the following sequence $\{K_n\}$ is exhaustive: $K_n = \{z : \text{dist } (z, \text{compl } G) \geq \frac{1}{n},\ |z| \leq n\}$.

Given an exhaustive sequence $\{K_n\}$, let

$$(3.3)\quad \rho(f,g) = \sum_{n=1}^{\infty} 2^{-n} \frac{\|f - g\|_{K_n}}{1 + \|f - g\|_{K_n}}$$

It is straightforward to verify that ρ is a metric on $C(G)$. To show that the metric topology is equivalent to the given one (in terms of the seminorms $\|\cdot\|_K$) we need only prove that a neighborhood of 0 in one topology contains a neighborhood of 0 in the other. This means that for each $\varepsilon > 0$, there is a compact set K and a $\delta > 0$ such that if $\|f\|_K < \delta$ then $\rho(f,0) < \varepsilon$; and conversely, for each compact K and $\delta > 0$ there is an $\varepsilon > 0$ such that if $\rho(f,0) < \varepsilon$ then $\|f\|_K < \delta$. This is left as an exercise for the reader.

The important characteristic of the topology we have imposed on $C(G)$ is the following proposition whose proof is trivial.

(3.4) Proposition: The sequence $\{f_n\}$ converges to f in $C(G)$ if and only if for each compact $K \subseteq G$, $f_n(z)$ converges to $f(z)$ uniformly for $z \in K$.

From this and well-known facts about uniform convergence we have:

(3.5) Proposition: $C(G)$ is complete.

Proof: If $\{f_n\}$ is a Cauchy sequence, then for each compact $K \subseteq G$, $f_n|_K$ is uniformly Cauchy on K and hence uniformly convergent on K to a function $f_{(K)}$ which is continuous on K. It is trivial to show that $f_{(K_1)}$ and $f_{(K_2)}$ agree on $K_1 \cap K_2$ and thus there is single function $f \in C(G)$ such that for each K, $f|_K = f_{(K)}$. This f is the the limit of the f_n. QED

(3.6) Proposition. $C(G)$ is separable. Consequently $H(G)$ is also separable.

Proof: The second statement follows from the first by Proposition 1.30. The first statement is a consequence of the Stone-Weierstrass Theorem. Let A denote the set of all polynomials in the two real variables x and y (where $z = x + iy$) with rational coefficients (i.e. the coefficients have real and imaginary parts that are rational). Then A is countable. Given a compact set $K \subset G$ the set $A|_K = \{p|_K : p \in A\}$ is a subalgebra of $C(K)$ satisfying the hypotheses of Theorem 2.29. Thus $A|_K$ is dense in $C(K)$. Let $K_1 \subset K_2 \subset \cdots$ be an exhaustive sequence in G. Let $\varphi \in C(G)$. For each n choose a polynomial $p_n \in A$ such that $\|\varphi - p_n\|_{K_n} < 1/n$. It is then easy to verify that $\rho(\varphi,p_n) \to 0$ as $n \to \infty$. Thus A is dense in $C(G)$. QED

The Cauchy theorem and Cauchy integral formula. We take the following point of view: we assume that the Cauchy theorem and integral formula for rectangles are known. Thus if f is holomorphic in $G \subseteq \mathbb{C}$

and if R is an open rectangle with $R^- \subseteq G$, then $\int_{\partial R} f = 0$, and if $z \in R$, then

$$f(z) = \frac{1}{2\pi i} \int_{\partial R} \frac{f(w)}{w - z}\, dw.$$

For holomorphic functions of several variables a similar formula holds. Consider a bi-rectangle $R = R_z \times R_w$ where R_z and R_w are rectangles in $\mathbb{C}$; we suppose that R^- lies in G, an open set in $\mathbb{C}^2$, and that f is holomorphic in G. Then

$$(3.7) \quad \int_{\partial^* R} f = 0$$

$$(3.8) \quad f(z,w) = \left(\frac{1}{2\pi i}\right)^2 \int_{\partial^* R} \frac{f(z',w')}{(z' - z)(w' - w)}\, dz'dw',$$

where $\partial^* R$ is the distinguished boundary of R, defined by $\partial^* R = \partial R_z \times \partial R_w$ and

$$\int_{\partial^* R} \varphi \quad \text{means} \quad \int_{\partial R_z} \left(\int_{\partial R_w} \varphi(z,w)\, dw \right) dz.$$

To prove (3.7), write

$$\int_{\partial^* R} f = \int_{\partial R_z} \left(\int_{\partial R_w} f(z,w)\, dw \right) dz$$

$$= \int_{\partial R_z} 0\, dz = 0,$$

since for each z, $f(z,w)$ is holomorphic in w. To prove (3.8) write for each $w \in R_w$

$$f(z,w) = \frac{1}{2\pi i} \int_{\partial R_z} \frac{f(z',w)}{z' - z}\, dz',$$

using the Cauchy integral formula in one variable. Now for each z', write

$$f(z',w) = \frac{1}{2\pi i} \int_{\partial R_w} \frac{f(z',w')}{w' - w}\, dw'.$$

Combining these two yields (3.8).

(3.9) Proposition. If f is holomorphic in $G \subseteq \mathbb{C}^2$ then $\partial f/\partial z$ is holomorphic in G.

Proof: If R is a bi-rectangle then

$$\frac{\partial f}{\partial z}(z,w) = \left(\frac{1}{2\pi i}\right)^2 \int_{\partial^* R} \frac{f(z',w')}{(z'-z)^2(w'-w)}\,dz'dw',$$

which follows on differentiating under the integral sign. From this formula it follows easily that $\partial f/\partial z$ is holomorphic. (Question: where did we use the continuity of f?) QED

(3.10) Definitions (Operator). If V and W are topological vector spaces, then a map $T : V \to W$ is an operator provided it is linear, that is $T(af + bg) = aT(f) + bT(g)$ for all $a, b \in \mathbb{C}$ and $f, g \in V$. We usually consider only continuous operators.

(3.11) (Bounded). We call a set A in a topological vector space E bounded provided that for each neighborhood N of 0, there is an $\varepsilon > 0$ such that $\varepsilon A \subseteq N$, where $\varepsilon A = \{\varepsilon a : a \in A\}$.

If the topology on E is determined by a family $\mathcal{P}$ of seminorms, it is easy to show that a set A is bounded if and only if for all $p \in \mathcal{P}$, $\sup\{p(a) : a \in A\}$ is finite.

(3.12) Definition. An operator T is compact if the closure of $T(A)$ is compact for each bounded set A.

(3.13) Proposition. Differentiation is a continuous operator on $\mathcal{H}(G)$.

By differentiation, we mean the operator D defined by $Df = f'$ in the one variable case and $Df = \partial f/\partial z$ (or $\partial f/\partial w$) in the two variable case.

Proof: We prove the one variable case; two variables is similar. D is certainly linear so it is enough to prove that if $f_n \to f$ then $f_n' \to f'$. Given a compact $K \subseteq G$, it must be shown that $f_n'(z)$ converges to $f'(z)$ uniformly for $z \in K$. By an open covering argument, it suffices to show that each $z_0 \in K$ has a neighborhood N_{z_0} in which $f_n' \to f'$ uniformly. Given z_0, let R be an open rectangle containing z_0 with $R^- \subseteq G$, and let N be a compact neighborhood of z_0 with $N \subseteq R$. Now for $z \in N$

$$f'(z) - f_n'(z) = \frac{1}{2\pi i}\int_{\partial R} \frac{f(w) - f_n(w)}{(w-z)^2}\,dw,$$

so

$$|f'(z) - f_n'(z)| \le \frac{1}{2\pi} \|f - f_n\|_R \frac{\text{length}(\partial R)}{[\text{dist}(N,\partial R)]^2}$$

Since $\|f - f_n\|_R \to 0$ independent of $z \in N$ the result follows. QED

(3.14) Proposition. $H(G)$ is a closed subspace of $C(G)$.

Proof: It is clear that $H(G)$ is a subspace. Suppose that $f_n \to f$, $f_n \in H(G)$. Let $z_0 \in G$ and take R and N as in the previous proof. Then for $z \in N$,

$$f_n(z) = \frac{1}{2\pi i} \int_{\partial R} \frac{f_n(w)}{w - z}\, dw.$$

Let $n \to \infty$ and use the uniform convergence to take the limit under the integral. This gives

$$f(z) = \frac{1}{2\pi i} \int_{\partial R} \frac{f(w)}{w - z}\, dw, \quad z \in N.$$

It follows that f is holomorphic in a neighborhood of z_0. The proof for $G \subseteq \mathbb{C}^2$ goes the same way. QED

NOTES: The theorem of Hartogs mentioned after Definition 3.1 may be found in [L. Hörmander].

Equation (3.3) is a standard method of converting the topology of a sequence of pseudo-metrics to that of a single metric. It occurs often in analysis. (See Chapter 17.)

Exercises

1. Let $\{K_n\}$ be a sequence of compact sets in the open set G such that $\cup K_n = G$ and $K_n \subseteq \text{int } K_{n+1}$ for each n. Show that $\{K_n\}$ is exhaustive in G. What happens if we replace $K_n \subseteq \text{int } K_{n+1}$ with only $K_n \subseteq K_{n+1}$?

2. Show that (3.3) defines a metric and that it induces a topology equivalent to the semi-norm topology on $C(G)$.

3. Show that if E is a topological vector space with a topology determined by a family P of seminorms, then a set $A \subseteq E$ is bounded if and only if for all $p \in P$, $\sup\{p(a) : a \in A\}$ is finite.

4. If $T : E \to F$ is a continuous operator between topological vector spaces and A is a bounded set in E, show that $T(A)$ is a bounded set in F.

If $S_1 : F \to V$ and $S_2 : W \to E$ are compact operators, show that $S_1T : E \to V$ and $TS_2 : W \to F$ are compact operators.

5. If $G = \mathbb{C}$ and $g \in H(G)$ show that $T : H(G) \to H(G)$ defined by $Tf = f \circ g$ is a continuous operator.

6. Justify the use of differentiation under the integral in Proposition 3.4.

7. For a function $f \in C(G)$ let $\text{supp } f = \{z \in G : f(z) \neq 0\}^-$ (the support of f). Let $C_c(G) = \{f \in C(G) : \text{supp } f$ is compact$\}$ and $C_0(G) = \{f \in C(G) : f(z) \to 0$ as $z \to \partial G$ or $|z| \to \infty\}$. Show that $C_c(G) \subseteq C_0(G) \subseteq C(G)$ and that $C_c(G)$ is dense in $C(G)$. Give $C_0(G)$ the topology induced by the single norm $\|f\|_G = \sup_{z \in G} |f(z)|$ and show that $C_c(G)$ is dense in $C_0(G)$.

8. Prove Liouville's Theorem (2.28) in n variables.

9. Suppose f and g are entire functions in n variables and that $f|_G = g|_G$ for some non-empty open set G. Prove that $f = g$. Suppose only that $f|_E = g|_E$ for a set E having a finite limit point. Can you conclude $f = g$?

§4. More About $C(G)$ and $H(G)$

It is commonly prove in courses of advanced calculus that compact sets in $\mathbb{R}$ (or more generally $\mathbb{R}^n$) are characterized by being closed and bounded. In a general topological vector space only one implication is correct. Here and in the future we will assume that our topological vector spaces are Hausdorff.

(4.1) Proposition: <u>In a topological vector space E each compact set is closed and bounded.</u>

Proof: Since we have assumed our topological vector spaces are Hausdorff, a compact set is automatically closed. Now let U be any neighborhood of the origin in E and $f \in E$. Continuity of scalar multiplication shows that there is a neighborhood V of the origin in E and an $\delta > 0$ s.t. $aV \subseteq U$ for all $a \in (-\delta,\delta)$. Replacing V by $\frac{1}{\delta}V$, we may suppose $\delta = 1$. Arguing again from continuity of scalar multiplication, we see that $\{nV : n = 1,2,3,\ldots\}$ is an open covering of E. If $Y \subseteq E$ is compact there is a finite set $n_1V, n_2V, \ldots, n_kV$ covering Y. If ε is the minimum of $\frac{1}{n_1},\ldots,\frac{1}{n_k}$ then $\varepsilon n_i V \subseteq U$, $i = 1,2,\ldots,k$, so $\varepsilon Y \subseteq U$ and Y is thus bounded. QED.

If U is a convex neighborhood of the origin then the introduction of V is unnecessary and the proof simplifies. In particular if the topology on E is given by a family of semi-norms then the result follows simply from the boundedness of continuous functions (the semi-norms) on compact sets. The following shows the divergence from the finite dimensional case.

(4.2) Proposition. <u>There is a closed and bounded set in $C(G)$ that is not compact.</u>

Proof: In the case of $C(G)$ (or $H(G)$) a set F is bounded if and only if for each compact $K \subseteq G$ the semi-norms $\|f\|_K$ are bounded as f runs over F. Choose any closed disk $K \subseteq G$ and let F consist of all $f \in C(G)$ that vanish outside of K and satisfy $\|f\|_K \le 1$. Then F is clearly closed and bounded. Consider the function $\varphi : F \to \mathbb{R}$ defined by

$$\varphi(f) = \frac{1}{\int_K 1\,dA - \int_K f\,dA}, \quad dA = dxdy.$$

A simple argument shows φ is continuous on F but unbounded and so F cannot be compact. QED

One of the striking properties of $H(G)$ is that the characterization of compact sets in $H(G)$ is formally the same as in $\mathbb{R}$.

(4.3) Main Theorem: Any subset of $H(G)$ that is closed and bounded must be compact.

Remark: This is the Stieltjes-Osgood Theorem. See the Notes at the end of this chapter.

Proof of the Theorem: Let F be a closed and bounded subset of $H(G)$. Since $H(G)$ is a metric space, it is enough to prove that any sequence $\{f_n\}$ in F has a subsequence $\{f_{n'}\}$ that converges in $H(G)$. Let K be a compact subset of G. We claim that F is a uniformly equicontinuous family of functions on K. Time out for a definition.

(4.4) Definition. A family F of functions on K is uniformly equicontinuous provided that for each $\varepsilon > 0$ there is a $\delta > 0$ such that if $|z - z'| < \delta$ and $z, z' \in K$, then $|f(z) - f(z')| < \varepsilon$ for each $f \in F$. (The important point is that δ does not depend on $f \in F$.)

To prove that F is uniformly equicontinuous let $\varepsilon > 0$ and let $\delta_1 = \frac{1}{2}\operatorname{dist}(K,\partial G)$. Let $K_1 = \{z \in \mathbb{C} : \operatorname{dist}(z,K) \leq \delta_1\}$. Then K_1 is compact and $K_1 \subseteq G$. Since F is bounded and differentiation is a continuous operator $F' = \{f' : f \in F\}$ is also bounded (Ex. 5, Ch. 3) and so $\|f'\|_{K_1}$ is bounded as f runs over F. If $z, z' \in K$ and $|z - z'| < \delta_1$ then $|f(z) - f(z')| = |\int_z^{z'} f'(\zeta)d\zeta| \leq \|f'\|_{K_1}|z - z'|$. Letting $\delta = \min(\delta_1, \varepsilon/\sup_f\|f'\|_{K_1})$ gives uniform equicontinuity on K.

Now choose a sequence of points $\{z_1,z_2,\ldots\}$ dense in K. Because $\{f_n\}$ is uniformly bounded on K, the sequence $\{f_n(z_1)\}$ is bounded (in $\mathbb{C}$!) and so has a convergent subsequence $\{f_{n,1}(z_1)\}$. Further, $\{f_{n,1}(z_2)\}$ has a convergent subsequence, say $\{f_{n,2}(z_2)\}$. (Note that $\{f_{n,2}(z_1)\}$ also converges.) Continuing, we get for each k, a subsequence $\{f_{n,k}\}$ of $\{f_n\}$ such that (i) $\{f_{n,k}(z_j)\}_{n=1}^{\infty}$ converges for $j = 1,2,\ldots,k$ and (ii) $\{f_{n,k}\}$ is a subsequence of $\{f_{n,k-1}\}$. Now we diagonalize, that is, look at the sequence $\{f_{n,n}\}$: it must converge at each point $z_1,z_2,z_3,\ldots$.

Finally, we have only to prove that a uniformly bounded, uniformly equicontinuous sequence $\{f_n\}$ that converges on a dense subset of K, converges uniformly on K. We need only show that $\{f_n\}$ is uniformly Cauchy. Let $\varepsilon > 0$ be given and let $\delta > 0$ be the number described in Definition 4.4. Let $z_1, z_2, \ldots, z_{m_0}$ be points chosen so that for each $z \in K$ there is a z_j with $j \leq m_0$ and $|z - z_j| < \delta$. Choose $n_0 = n_0(\varepsilon)$ so that $m,n \geq n_0$ implies that $|f_m(z_j) - f_n(z_j)| < \varepsilon$ for $j = 1,2,\ldots,m_0$. Then for any $z \in K$ we have, if $m,n \geq n_0$,

$$|f_m(z) - f_n(z)| \leq |f_m(z) - f_m(z_j)| +$$

$$|f_m(z_j) - f_n(z_j)| + |f_n(z_j) - f_n(z)|$$

$$\leq \varepsilon + \varepsilon + \varepsilon = 3\varepsilon,$$

where z_j is chosen so that $|z - z_j| < \delta$, $j \leq m_0$. Hence $\{f_n\}$ is uniformly Cauchy and so converges on K.

Repeat this procedure for an exhaustive sequence $K_1, K_2, K_3, \ldots$ of compact subsets of G with $K_1 \subseteq K_2 \subseteq \cdots$. (i.e find a subsequence converging on K_1 and a subsequence of that subsequence converging on K_2, etc.) Diagonalize again to obtain a subsequence converging in $\mathcal{H}(G)$. Similar arguments work in the case of several variables. QED

(4.5) Corollary: The identity operator on $\mathcal{H}(G)$ is compact. In fact any continuous operator into $\mathcal{H}(G)$ is compact.

Proof: The first statement is trivial and the second follows from Ex. 5, Ch. 3 on composing with the identity. QED

(4.6) The main theorem above is enormously useful in complex function theory. It will be used later in our proof of the Riemann Mapping Theorem. It is nearly always the main step in proving spaces of analytic functions are complete. The enterprising students can try their hands at proving that the following normed space is complete: Let $G = \mathbb{D}$ be the unit disk $\{z : |z| < 1\}$ and define

$$B_1 = \{f \in \mathcal{H}(\mathbb{D}) : \int_0^{2\pi} \int_0^1 |f(re^{i\theta})| r\,dr\,d\theta < +\infty\}.$$

B_1 is topologized by the single norm $\|f\|_1 = \int_0^{2\pi} \int_0^1 |f(re^{i\theta})| r\,dr\,d\theta$ which renders it a metric space. The first step is to show that the ball $\{f \in B_1 : \|f\|_1 \leq 1\}$ is bounded in $\mathcal{H}(\mathbb{D})$, i.e. relative to the semi-norms $\|\cdot\|_K$.

An easier example is the space

$$A_{-1} = \{f \in H(\mathbb{D}) : \text{There exists a constant } C \text{ with } |f(z)| \leq C(1 - |z|)^{-1}, \text{ all } z \in \mathbb{D}\}.$$

A_{-1} is given the norm $\|f\| = \sup\{|f(z)|(1 - |z|) : z \in D\}$. (The subscripts 1 and -1 in these examples refer to the exponent of $|f|$ and of $1 - |z|$, respectively, in the definition of these spaces.)

Taking this second example, it follows easily that among all functions $f \in H(\mathbb{D})$ with $|f(z)| \leq (1 - |z|)^{-1}$ there is one which maximizes $\dfrac{\text{Re } f''(0)}{1 + |f(\frac{1}{2})|^2}$. This is really a silly example and it is easy to manufacture others like it. To prove the assertion, let $\varphi(f)$ denote the expression above. Then φ is continuous on $H(\mathbb{D})$ and so must attain its maximum on the compact set F described by the condition $|f(z)| \leq (1 - |z|)^{-1}$.

A more interesting application is obtained by letting $\varphi(f) = \int_0^{2\pi} \int_0^{\frac{1}{2}} |f'(re^{i\theta})|^2 r dr d\theta$. Then $\varphi(f)$ is the area of the image of the disk $D' = \{z : |z| < \frac{1}{2}\}$ under f (counting overlaps). It follows that among all functions $f \in H(\mathbb{D})$ such that $|f(z)| \leq 1$ for all $z \in \mathbb{D}$, there is at least one that maps D' onto a set of maximum area.

NOTES: The reader familiar with topology will recognize that $C(G)$ carries the <u>compact-open</u> topology and so there exists a characterization of the compact subsets of $C(G)$ in which the notion of equicontinuity plays a crucial role. This is the Arzela-Ascoli Theorem. See, for example [Dugundgi, p. 267], and [Kelley, p. 233].

The Main Theorem is the Stieltjes-Osgood Theorem [Saks-Zygmund, p. 119] on <u>normal families</u>. A family $F \subseteq H(G)$ is called normal if for any sequence $\{f_n\}$ in F either there is a subsequence converging uniformly on compact subsets of G to a holomorphic function or else there is a subsequence converging uniformly on compact subsets to ∞. The Stieltjes-Osgood Theorem says that any family F such that $\sup\{|f(z)| : z \in G\} \leq M$ for all $f \in F$ is a normal family when $M < +\infty$. Of course the boundedness condition rules out the option of converging to ∞ and so not only is F normal but also compact.

Exercises

1. Show that the function φ defined in Proposition 4.2 is continuous and unbounded on $\mathcal{F}$.

2. Show that B_1 and A_{-1} defined in 4.6 are complete metric spaces.

3. Imitate the argument in the Main Theorem (4.3) to show that a closed bounded set in $C(G)$, which is uniformly equicontinuous on each compact subset of G, must be compact. Deduce the following: Let $\mathcal{F}$ be closed and bounded in $C(G)$ such that every $f \in \mathcal{F}$ has continuous first partial derivitives and $\{\partial f/\partial x, \partial f/\partial y : f \in \mathcal{F}\}$ is bounded. Then $\mathcal{F}$ is compact.

4. Let $\mathcal{F} \subseteq C(G)$ and $z_0 \in G$. $\mathcal{F}$ is said to be <u>equicontinuous at</u> z_0 if for any $\varepsilon > 0$ there is a $\delta > 0$ such that for any $f \in \mathcal{F}$ and any z with $|z - z_0| < \delta$, we have $|f(z) - f(z_0)| < \varepsilon$. $\mathcal{F}$ is said to be equicontinuous on $A \subseteq G$ if $\mathcal{F}$ is equicontinuous at every point in A. Show that if $K \subseteq G$, K compact, then $\mathcal{F}$ is equicontinuous on K if and only if it is uniformly equicontinuous on K.

5. The unit ball $\{f \in A_{-1} : \|f\| \leq 1\}$ is compact in $H(\mathbb{D})$. Show, however, that it is not compact in A_{-1} with the topology given by the norm $\|\cdot\|$. (See (4.6).)

6. For $f \in H(\mathbb{D})$ let $a_n(f)$ be the nth coefficient of its power series expansion about zero. Show that a set $\mathcal{F} \subseteq H(\mathbb{D})$ is compact if and only if it is closed and for each $n, \{a_n(f) : f \in \mathcal{F}\}$ is bounded and $\limsup b_n^{1/n} \leq 1$ where $b_n = \sup_{f\in\mathcal{F}} |a_n(f)|$.

7. Characterize compactness for subsets of $H(\mathbb{C})$ in terms of the power series coefficients. (See Exercise 6.)

8. Show that $\mathcal{F} \subseteq H(G)$ is compact if and only if each $\zeta \in G$ is contained in a rectangle $R_\zeta \subseteq G$ such that

$$\sup_{f\in\mathcal{F}} \iint_{R_\zeta} |f(z)|\,dxdy < +\infty$$

9. For $\mathcal{F} \subseteq H(G)$ let $\mathcal{F}' = \{f' : f \in \mathcal{F}\}$. Assume G is connected and show that $\mathcal{F}$ is bounded in $H(G)$ whenever $\mathcal{F}'$ is bounded and for some point $z_0 \in G$, $\{f(z_0) : f \in \mathcal{F}\}$ is bounded.

§5. Duality

The concept of duality is one of the most productive in analysis. One can very nearly characterize applications of functional analysis as applications of duality. Most of our goal in succeeding sections will be identifying the dual of $H(G)$ and exploiting that identification. Toward this end, we begin with the definition.

(5.1) Definition. If E is a vector space over $\mathbb{C}$, then a linear functional is a mapping $L : E \to \mathbb{C}$ such that $L(af + bg) = aL(f) + bL(g)$ for $a, b \in \mathbb{C}$ and $f, g \in E$. When E is a topological vector space and L is continuous, L is called a continuous linear functional. The collection of continuous linear functionals on E, written E^*, is called the dual of E.

The point of studying E^* is that its structure gives much information about the structure of E, and often in a useful way. Ideally, if E has a concrete representation, one would hope for a concrete representation of E^*. For instance, a finite dimensional space E can be represented via bases as $\mathbb{C}^n$ and similarly E^* can be represented as $\mathbb{C}^n$. The representations are such that if $L \in E^*$ is represented by $(\lambda_1,\ldots,\lambda_n)$ and $f \in E$ is represented as $(z_1,\ldots,z_n)$ then $L(f) = \sum_{j=1}^{n} \lambda_j z_j$. (There are other modes of representation but this seems easiest to derive.) Thus many problems in finite dimensional spaces are reduced to purely computational ones. It turns out $H(G)$ has a nice concrete representation which has important applications.

Linear functionals are the natural objects to study on vector spaces. If the space has a multiplicative structure then the following are useful as well.

(5.2) Definition. If E is a vector space which also happens to be an algebra, then a multiplicative linear functional (mlf) is an algebra homomorphism of E into $\mathbb{C}$. That is, an mlf is a linear functional which has the further property: $L(fg) = L(f)L(g)$, $f, g \in E$. We exclude $L \equiv 0$.

The next result tells us what the mlf's are for $H(G)$.

(5.3) Theorem. On $H(G)$, the mlf's are just the point evaluations.

That is, if L is an mlf, then there is a $z_0 \in G$ such that $L(f) = f(z_0)$ for all $f \in H(G)$. Of course, each point evaluation is an mlf.

Proof: If $c \in \mathbb{C}$, let us use $\tilde{c}$ to denote the constant function $\tilde{c}(z) = c$. Let L be an mlf. Now $L(\tilde{1}) = L(\tilde{1} \cdot \tilde{1}) = L(\tilde{1})L(\tilde{1})$ so $L(\tilde{1})$ is either 0 or 1. If $L(\tilde{1}) = 0$ then $L(f) = L(\tilde{1} \cdot f) = 0 \cdot L(f) = 0$ for every $f \in H(G)$. Excluding the identically zero functional, we must have $L(\tilde{1}) = 1$. Then $L(\tilde{c}) = L(c \cdot \tilde{1}) = cL(\tilde{1}) = c$. Now let $\hat{z}$ denote the identity function: $\hat{z}(z) = z$ for each $z \in G$.

Claim: If $z_0 = L(\hat{z})$ then $z_0 \in G$. For if $z_0 \notin G$ then $\frac{1}{z - z_0} \in H(G)$. Put another way, there is an $f \in H(G)$ such that $(\hat{z} - \tilde{z}_0)f = \tilde{1}$. Thus, on the one hand $L(\tilde{1}) = 1$ but on the other $L((\hat{z} - \tilde{z}_0)f) = (z_0 - z_0)L(f) = 0$. This contradiction means z_0 must lie in G. Now, for any $f \in H(G)$ the function $\frac{f(z) - f(z_0)}{z - z_0}$ --defined to be $f'(z_0)$ when $z = z_0$ --also lies in $H(G)$. This means there is a $g \in H(G)$ such that $(\hat{z} - \tilde{z}_0)g = f - f(z_0)^{\sim}$. Applying L to both sides of this gives $0 = L(f) - f(z_0)$. QED

We will see later that this theorem need not hold for certain open sets in $\mathbb{C}^n$, $n \geq 2$.

A homomorphism on $H(G)$ with range in $\mathbb{C}$ is an mlf and has a simple representation, as we have just seen. A homomorphism with range in $H(G')$ for another open set $G' \subseteq \mathbb{C}$ also has a simple representation. First, observe that if $\varphi : G' \to G$ is holomorphic then $\alpha : H(G) \to H(G')$ defined by $\alpha(f) = f \circ \varphi$ is an algebra homomorphism. The converse to this is true if G' is connected.

(5.4) Proposition: Suppose α is any non-zero algebra homomorphism from $H(G)$ to $H(G')$ where G' is connected. Then there is a holomorphic map $\psi : G' \to G$ such that $\alpha(f) = f \circ \psi$ for all $f \in H(G)$.

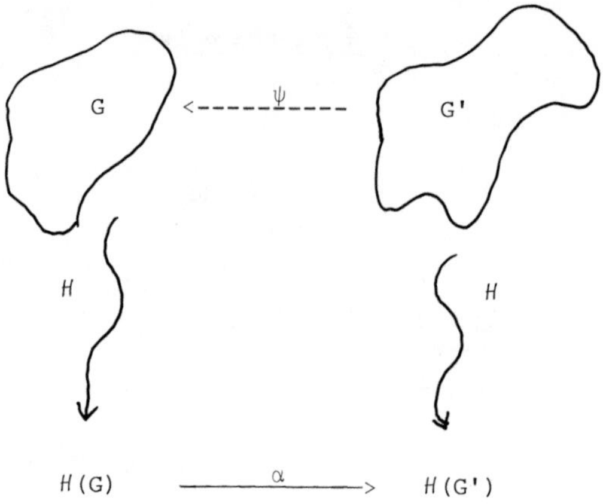

Figure 5.1

Proof: From $\alpha(\tilde{1}) = \alpha(\tilde{1})\alpha(\tilde{1})$ we again see that $\alpha(\tilde{1}) = \tilde{1}$. For $\alpha(\tilde{1})$ must be a function whose range is in $\{0,1\}$. Since G' is connected and $\alpha(\tilde{1})$ is continuous on G', $\alpha(\tilde{1})$ must be constant. Since $\alpha \equiv 0$ is excluded $\alpha(\tilde{1})$ must be identically one. Also $\alpha(\tilde{c}) = \tilde{c}$ for any $c \in \mathbb{C}$. Now for any $z' \in G'$ define $L_{z'} : H(G) \to \mathbb{C}$ by $L_{z'}(f) = \alpha(f)(z')$. That is, $L_{z'}$ is the composition of evaluation at z' with α. Clearly $L_{z'}$ is a mlf on $H(G)$ and so, by the preceding theorem, there is a $z_0 \in G$ such that $L_{z'}(f) = f(z_0)$. Taking $f = \hat{z}$ gives $z_0 = L_{z'}(\hat{z}) = \alpha(\hat{z})(z')$. Thus for arbitrary $z' \in G'$,

$$\alpha(f)(z') = L_{z'}(f) = f(z_0) = f(\alpha(\hat{z})(z')),$$

which says $\alpha(f) = f \circ \psi$ with $\psi = \alpha(\hat{z})$. QED

Since Theorem 5.3 is used in proving Proposition 5.4 and is, moreover, a special case of the proposition (take α with only constants in its range) it is clear that Proposition 5.4 also need not hold in several variables. Now we prove the failure of Theorem 5.3 in two (or more) variables.

(5.5) Proposition: Let R_1 and R_2 be open rectangles each containing 0, and suppose $f(z,w)$ is holomorphic in the open set $R_1 \times R_2 \setminus \{(0,0)\}$. Then f is holomorphic at $(0,0)$ as well. This means that there is a function φ holomorphic in $R_1 \times R_2$ such that $f(z,w) = \varphi(z,w)$ for $(z,w) \in R_1 \times R_2 \setminus \{(0,0)\}$.

Suppose this is granted. Then we can define a mlf L on $H(R_1 \times R_2 \backslash \{(0,0)\})$ by setting $L(f) = \varphi(0,0)$. It is clear that f corresponds to a unique φ and this makes it a routine exercise to show L is an mlf. (Roughly speaking we have $H(R_1 \times R_2 \backslash \{(0,0)\}) = H(R_1 \times R_2)$.) Now there can be no $(z_0,w_0) \in R_1 \times R_2 \backslash \{(0,0)\}$ such that $L(f) = f(z_0,w_0)$. [Proof: Consider functions $f(z,w) = z - z_0$ and $g(z,w) = w - w_0$. $L(f) = -z_0$ and $L(g) = -w_0$ but $f(z_0,w_0) = 0$ and $g(z_0 \cdot w_0) = 0$. Thus L cannot be evaluation at (z_0,w_0) for $(z_0,w_0) \neq (0,0)$.]

Proof of Proposition: Let R_2' be an open rectangle containing 0 whose closure lies in R_2. Let $\varphi(z,w) = \frac{1}{2\pi i} \int_{\partial R_2'} \frac{f(z,\zeta)}{\zeta - w}\, d\zeta$. Then φ is holomorphic in $R_1 \times R_2'$. But as long as $z \neq 0$ the Cauchy integral formula tells us $\varphi(z,w) = f(z,w)$. By continuity, $\varphi(z,w) = f(z,w)$ in $R_1 \times R_2'$, so long as $(z,w) \neq (0,0)$. Define φ outside $R_1 \times R_2'$ in $R_1 \times R_2$ by setting it equal to f and we are done. QED

Let $G = \mathbb{D}$ be the unit disk $\{z : |z| < 1\}$. Every function $f \in H(\mathbb{D})$ has a power series expansion $f(z) = \sum_{n=0}^{\infty} a_n z^n$, $z \in D$. Thus, elements of $H(\mathbb{D})$ can be represented by sequences, whereby each function is represented by its Taylor coefficients. Not every sequence corresponds to an analytic function since the root test for convergence implies $\limsup |a_n|^{1/n} \leq 1$ whenever $\sum_{n=0}^{\infty} a_n z^n$ converges in $\mathbb{D}$. We will investigate some consequences of this representation in the next section. For use in that section we prove the following result.

(5.6) Proposition: Suppose $\limsup |b_n|^{1/n} < 1$. Then $\sum a_n b_n$ converges for each sequence $\{a_n\}$ satisfying $\limsup |a_n|^{1/n} \leq 1$. Conversely, if $\{b_n\}$ has this last property, then $\limsup |b_n|^{1/n} < 1$.

Proof: If $\limsup |b_n|^{1/n} < 1$ then $\limsup |a_n b_n|^{1/n} < 1$ as well, and so $\sum a_n b_n$ converges by the root test. On the other hand, if $\limsup |b_n|^{1/n} \geq 1$ then there is a sequence $n_k \to \infty$ and a sequence $\varepsilon_k \to 0$ such that $|b_{n_k}| \geq (1 - \varepsilon_k)^{n_k}$. Define a_n by (a) $a_n = 0$ unless n is an n_k, (b) $|a_{n_k}| = (1 - \varepsilon_k)^{-n_k}$ and (c) $a_n b_n \geq 0$. Then $\limsup |a_n|^{1/n} = 1$, but $\sum a_n b_n = \infty$. This proves the converse.

NOTES: The concept of a multiplicative linear functional makes sense for any algebra over $\mathbb{C}$ and is connected with the concept of maximal ideals. (See Chapter 13 on ideals in $H(G)$.)

The property of $H(G)$ proved in Theorem 5.3 is shared by $C(G)$, that is every mlf on $C(G)$ is given by evaluation at a point of G. The proof of this is beyond the scope of this book. See [Dugundji, Ch. XIII, Sect. 6].

A theorem more general than Proposition 5.5 is due to Hartogs. A proof may be found in [Hörmander, p. 30].

The notion of duality is common to all books in functional analysis.

Exercises

1. Describe the algebra homomorphisms α from $H(G)$ to $H(G')$ if G' is not connected. (What are the possibilities for $\alpha(\tilde{1})$?)

2. Prove the following stronger form of Proposition (5.5): Let R_1 and R_2 be open rectangles and F_i a compact subset of R_i, $i = 1,2$. Suppose $f(z,w)$ is holomorphic in $(R_1 \setminus F_1) \times (R_2 \setminus F_2)$ Then there is a function $\varphi(z,w)$ holomorphic in $R_1 \times R_2$ such that $f(z,w) = \varphi(z,w)$ for $(z,w) \in (R_1 \setminus F_1) \times (R_2 \setminus F_2)$.

3. Show that every mlf on $H(G)$ is continuous. Show, in fact, that every algebra homomorphism $\alpha : H(G) \to H(G')$ is continuous.

4. Let g be a continuous function on the open set G such that $\{z \in G : g(z) \neq 0\}^-$ is compact. Show that $Tf = \int_G fg\,dxdy$ defines a continuous linear functional on $C(G)$ but T is not an mlf.

5. If A is an algebra with a multiplicative identity 1 and if $f \in A$ then the <u>spectrum</u> of f (written spec f) is the collection of $\lambda \in \mathbb{C}$ such that $f - \lambda \cdot 1$ does not have an inverse in A (i.e. the equation $fg - \lambda g = 1$ has no solution g in A.) Show that for $f \in H(G)$, spec $f = f(G) = \{\alpha(f) : \alpha$ is an mlf on $H(G)\}$.

Let $A = H^\infty(G) \equiv \{f \in H(G) : f$ is bounded$\}$. Show that spec $f = f(G)^-$.

6. Let $G = \mathbb{D}$ (the unit disk) and $G' = H$ (the right half-plane Re $z > 0$.) Show that $H(G)$ and $H(G')$ are isomorphic. (Topological algebras are isomorphic if there is a continuous one-to-one, onto homomorphism from one to the other whose inverse is continuous.)

7. Let $z_0 \in G$ and let $L : H(G) \to \mathbb{C}$ be a continuous linear functional. L is called a <u>point derivation at</u> z_0 if $L(fg) = f(z_0)L(g) + g(z_0)L(f)$. Show that there is a constant C such that $L(f) = Cf'(z_0)$.

8. Let L be a linear functional on $H(G)$ such that $|L(f)| \leq \sup\{|f'(z)| : z \in K\}$ for a certain compact set K. Show that L is continuous on $H(G)$.

9. Use the characterization of bounded in Exercise 6 of Chapter 4 to show that if $F \subseteq H(\mathbb{D})$ is bounded and L is defined on $H(\mathbb{D})$ by

$$L(f) = \sum a_n(f)b_n$$

where $\limsup |b_n|^{1/n} < 1$, then $\{L(f) : f \in F\}$ is bounded.

10. Let E be a topological vector space, F a bounded subset of E and L a linear functional on E which is not bounded on F. Show that L is not continuous.

§6. Duality of *H*(G)—The Case of the Unit Disc

We begin with a general result about linear functionals on a locally convex topological vector space. Let E have the topology generated by a family P of seminorms. For each non-empty finite set $A = \{\|\cdot\|_1, \|\cdot\|_2, \ldots, \|\cdot\|_n\} \subset P$, define $\|x\|_A = \max_{1\le j\le n} \|x\|_j$, $x \in E$. Then $\|\cdot\|_A$ is a seminorm. Let $\tilde{P} = P \cup \{\|\cdot\|_A : A$ is a non empty finite subset of $P\}$; then P and $\tilde{P}$ generate the same topology on E (Exercise 2). Consequently, we may assume $P = \tilde{P}$ in the following proposition.

(6.1) Proposition: A linear functional L on a locally convex topological vector space E is continuous if and only if there is a seminorm $\|\cdot\|_0$ and a number A such that $|L(f)| \le A\|f\|_0$ for all $f \in E$.

Proof: First suppose $|L(f)| \le A\|f\|_0$. Then $\|f\| < \varepsilon/A$ implies $|L(f)| < \varepsilon$ so that L is continuous at 0 and hence, by linearity, everywhere.

Conversely, suppose L is continuous. Then $L^{-1}(U)$ is open if $U = \{z : |z| < 1\}$. Therefore, there is a neighborhood N of 0 such that $f \in N$ implies $|L(f)| < 1$. Now, there is a seminorm $\|\cdot\|_0$ and an $\varepsilon > 0$ such that $\|f\|_0 \le \varepsilon$ implies $f \in N$. (This requires the assumption discussed above.) Let $f \in E$ be such that $\|f\|_0 \ne 0$. Then $\varepsilon f/\|f\|_0 \in N$ which implies $|L(\varepsilon f/\|f\|_0)| < 1$ or $|L(f)| < 1/\varepsilon\|f\|_0$. If $\|f\|_0 = 0$, then $nf \in N$ for all integers $n > 0$ so $|L(nf)| < 1$. This clearly means $L(f) = 0$. Thus, in either case, $|L(f)| \le A\|f\|_0$ with $A = 1/\varepsilon$. QED

(6.2) Our problem is to represent the dual of $H(G)$. Here we treat the special case $G = \mathbb{D} = \{z : |z| < 1\}$. The solution in this simple case gives us some insight into the general problem. We will use the fact, whose proof requires the Cauchy integral formula for disks, that if $f \in H(\mathbb{D})$, then the Taylor series for f, $\sum a_n z^n$, where $a_n = f^{(n)}(0)/n!$, converges to f in the topology of $H(\mathbb{D})$. The analogous results hold for functions of several variables: if $f \in H(\mathbb{D} \times \mathbb{D})$ then the Taylor series of f, $\sum a_{mn} z^m w^n$, $a_{mn} = \frac{1}{m!n!}\frac{\partial^{m+n} f}{\partial z^m \partial w^n}$, converges to f in the topology of $H(\mathbb{D} \times \mathbb{D})$. This means the following. For any compact set K in $\mathbb{D} \times \mathbb{D}$ and any $\varepsilon > 0$, there exist N_0, M_0

such that $\| f(z,w) - \sum_{m=0}^{M} \sum_{n=0}^{N} a_{mn} z^m w^n \|_K < \varepsilon$ whenever $M > M_0$, $N > N_0$, in $H(\mathbb{D} \times \mathbb{D})$. There is no problem in generalizing the results of this section to the case of several variables. This we leave as an exercise for the interested reader.

We identify $H(\mathbb{D})$ with the space A of sequences $a = \{a_n\}$ satisfying $\sigma(a) \le 1$, where $\sigma(a) = \lim\sup |a_n|^{1/n}$. Explicitly, consider the map $\varphi : A \to H(\mathbb{D})$ defined by

$$\varphi(a) = f|_{\mathbb{D}}$$

where $f(z) = \sum a_n z^n$. This mapping is linear and has an inverse ψ given by $\psi(f) = \{f^{(n)}(0)/n!\}$. This correspondence gives us a topology on A. We need a more direct representation of the topology on A.

(6.3) Definition: The locally convex topological vector space A is the above sequence space with the topology given by the seminorms $\|a\|_r$, $0 < r < 1$, where $\|a\|_r = \sup_n |a_n| r^n$.

We have to show that this gives the same topology. We must show that for each $\varepsilon > 0$ and $0 < r < 1$, there is a compact set $K \subseteq \mathbb{D}$ and a number $\delta > 0$ such that $\|f\|_K < \delta$ implies $\|a\|_r < \varepsilon$, where $f = \varphi(a)$. In the other direction we have a similar statement to show.

Choose $K = \{z : |z| \le r\}$. If $\|f\|_K < \delta$ then by Cauchy's inequalities (2.14) we have $|a_n| \le \delta/r^n$, i.e. $\|a\|_r \le \delta$ so we may take $\delta = \varepsilon$. In the other direction, suppose we are given $K \subseteq \mathbb{D}$, K compact, and $\varepsilon > 0$. Choose $\rho < 1$ so that $K \subseteq \{z : |z| < \rho\}$ and then choose r with $\rho < r < 1$. If we let $\delta = \varepsilon(1 - \rho/r)$ then $\|a\|_r < \delta$ implies

$$\begin{aligned} \|f\|_K &= \|\textstyle\sum a_n z^n\|_K < \sum |a_n| \rho^n \\ &= \textstyle\sum |a_n| r^n \frac{\rho^n}{r^n} \le \|a\|_r \sum \left(\frac{\rho}{r}\right)^n \\ &< \frac{\delta}{1 - \rho/r} = \varepsilon \end{aligned}$$

which is what we needed.

(6.4) Now, given $L \in H(\mathbb{D})^*$, let $\lambda_n = L(z^n)$, $n = 0,1,2,\dots$. Then, since $\sum_{n=0}^{m} a_n z^n \to f$ in $H(\mathbb{D})$ as $m \to \infty$, and since L is continuous $H(\mathbb{D})$, we have $L\left(\sum_{n=0}^{m} a_n z^n\right) \to L(f)$. But, as $L\left(\sum_{n=0}^{m} a_n z^n\right) = \sum_{n=0}^{m} a_n \lambda_n$, it follows that $\sum a_n \lambda_n$ converges and equals $L(f)$. From 5.6 we know that $\sigma(\lambda) < 1$. In the opposite direction, suppose $\lambda = \{\lambda_n\}$ is given with $\sigma(\lambda) < 1$. Define L by $L(f) = \sum a_n \lambda_n$ which converges, again by 5.6. It is clear that L is a linear functional, so to show it is in $H(\mathbb{D})^*$ we have only to find an r, $0 < r < 1$, and a number A such that $|L(f)| \leq A\|a\|_r$. Choose r and ρ such that $\sigma(\lambda) < \rho < r < 1$. Then there is an integer $n_0 > 0$ such that for $n \geq n_0$ we have $|\lambda_n| \leq \rho^n$. Then

$$|L(f)| = |\sum a_n \lambda_n| \leq \sum_{n<n_0} |a_n||\lambda_n| + \sum_{n \geq n_0} |a_n|\rho^n$$

$$\leq \|a\|_r \left(\sum_{n<n_0} \left|\frac{\lambda_n}{r^n}\right| + \frac{\rho^{n_0}/r^{n_0}}{1 - \rho/r} \right).$$

We have therefore proved the next result:

(6.5) Proposition: The dual of A is the space of all sequences λ with $\sigma(\lambda) < 1$.

With a minor abuse of language already abused above, we may restate this.

(6.6) Proposition. The dual of $H(\mathbb{D})$ is the space of all sequences λ with $\sigma(\lambda) < 1$.

Remark: In at least one direction the proof above could have been simplified by using seminorms $\|a\|_r' = \sum |a_n| r^n$ rather than $\|a\|_r = \sup_n |a_n| r^n$. It is not hard to show these are equivalent (generate the same topology).

The temptation is large, given a sequence λ such that $\sigma(\lambda) < 1$, to form the "generating function" $\sum \lambda_n z^n$. We will succumb to this temptation, but in a modified form.

(6.7) Definition: Given a set A in $\mathbb{C}^\wedge$, not necessarily open, to say that a function F is holomorphic on A means there is an open set $B \supseteq A$ such that F is holomorphic in B.

Note that functions holomorphic on A have domains that are open supersets of A. If A is open we may take $B = A$ and there is no confusion with the usual meaning of holomorphic.

(6.8) Definition: Given F_1 and F_2 holomorphic on A, we write $F_1 \sim F_2$ to mean that there is an open set $B \supset A$ such that both F_1 and F_2 are holomorphic on B and $F_1(z) = F_2(z)$ for $z \in B$.

As the notation implies, this is an equivalence relation between functions holomorphic on A (trivial) and we will write $[F]$ for the equivalence class that F belongs to. Sometimes, when confusion is unlikely, we simply write F instead of $[F]$.

Example: Let A be the singleton $\{0\}$. Notice that for $F_1 \sim F_2$ it is not enough to have $F_1(0) = F_2(0)$: Take $F_1(z) = z$, $F_2(z) = z^2$, then $F_1(0) = F_2(0)$ but F_1 and F_2 do not agree in any open set containing 0. Any F analytic at 0 has a power series $F(z) = \sum a_n z^n$ which converges in some neighborhood of 0; that is $\limsup |a_n|^{1/n} < +\infty$. It is then easy to show that $[F]$ is determined by the sequence $\{a_n\}$.

(6.9) Definition: The equivalence classes under $\sim$ are called the germs of holomorphic functions on A. $H(A)$ denotes the space of all germs of holomorphic functions on A.

It is clear how to add, multiply and differentiate a germ of a holomorphic function: simply do these things to representative functions in a suitable neighborhood of A.

It is possible to put a topology on $H(A)$. For the time being, at least, we will not do this. See Chapters 17 and 18.

(6.10) Definition: If $\infty \in A$, then $H_0(A)$ is the subspace of $H(A)$ consisting of all $[F]$ with $F(\infty) = 0$.

These definitions allow us to give another characterization of the dual of $H(\mathbb{D})$ that is more easily generalized.

(6.11) Theorem: $H(\mathbb{D})^* = H_0(\hat{\mathbb{C}} \setminus \mathbb{D})$. More precisely, given $F \in H_0(\hat{\mathbb{C}} \setminus \mathbb{D})$, let $L_F \in H(\mathbb{D})^*$ be defined by

$$L_F(f) = \frac{1}{2\pi i} \int_\Gamma f(z)F(z)\,dz,$$

where Γ is a circle in $\mathbb{D}$ such that F is holomorphic on and outside of Γ. Then, for each $L \in H(\mathbb{D})^*$ there is an $F \in H_0(\hat{\mathbb{C}} \setminus \mathbb{D})$ such that $L = L_F$ and this F satisfies

$$F(w) = L\left(\frac{1}{z - w}\right) \quad \text{for } w \notin \mathbb{D}.$$

Proof: It is clear that L_F is in $H(\mathbb{D})^*$, since L_F is linear and $|L_F(f)| \leq \|f\|_\Gamma \|F\|_\Gamma$. To show that each $L \in H(\mathbb{D})^*$ arises from an F in this manner, let $\lambda = \{\lambda_n\}$ be the associated sequence $\lambda_n = L(z^n)$. Let F be defined by $F(z) = \sum \lambda_n/z^{n+1}$. Because $\sigma(\lambda) < 1$, this series converges in a neighborhood of $\mathbb{C}^\wedge\backslash\mathbb{D}$ and so defines an element of $H_0(\mathbb{C}^\wedge\backslash\mathbb{D})$ We must now show that $L = L_F$. It is enough to prove that $L(z^n) = L_F(z^n)$ for $n = 0,1,2,\ldots$. But

$$L_F(z^n) = \frac{1}{2\pi i}\int_\Gamma z^n F(z)\,dz$$

$$= \sum_{k=0}^{\infty} \lambda_k \frac{1}{2\pi i}\int_\Gamma \frac{z^n}{z^{k+1}}\,dz = \lambda_n$$

where Γ is any circle in $\mathbb{D}$ which lies in the domain of convergence of $\sum \lambda_n/z^{n+1}$, i.e. in $\{z : |z| > \sigma(\lambda)\}$. This, incidentally, shows that the definition of L_F is independent of the choice of Γ.

Finally, for $|w| \geq 1$ and $|z| < 1$, $\frac{1}{w-z} = \sum_{n=0}^{\infty} \frac{z^n}{w^{n+1}}$ with convergence in $H(\mathbb{D})$. Thus $L(\frac{1}{w-z}) = \sum L(z^n)/w^{m+1} = \sum \lambda_n/w^{n+1} = F(w)$, concluding the proof of the theorem. QED

This theorem can be generalized to arbitrary domains, a task we postpone until we have had a chance to apply some of the results so far. However, it can also be generalized to special domains in an essentially trivial manner. We illustrate by sketching one such generalization.

(6.12) Proposition: If G is a finite union of disjoint disks then $H(G)^* = H_0(\mathbb{C}^\wedge\backslash G)$.

Proof. (Sketch). Say $G = D^1 \cup\cdots\cup D^n$ and let $L \in H(G)^*$. Now, elements of $H(G)$ are just n-tuples of functions holomorphic each in the appropriate disk. There are germs $F_j \in H_0(\mathbb{C}^\wedge\backslash D^j)$ and circles $\Gamma_j \subseteq D^j$, $j = 1,2,\ldots,n$ such that

$$L(f) = \sum_{j=1}^{n} \frac{1}{2\pi i}\int_{\Gamma_j} f(z)F_j(z)\,dz.$$

But, if $i \neq j$, then $\int_{\Gamma_j} f(z)F_i(z)\,dz = 0$. Thus we may write

$$L(f) = \sum_{j=1}^{n} \frac{1}{2\pi i}\int_{\Gamma_j} f(z)F(z)\,dz,$$

where $F = F_1 + F_2 + \cdots + F_n \in H_0(\hat{\mathbb{C}} \setminus G)$. The correspondence $L \leftrightarrow F$ identifies $H(G)^*$ with $H_0(\hat{\mathbb{C}} \setminus G)$.

NOTES: The study of duality in relation to $H(G)$ has been around for some time. Some of the earlier works are [Sebastiao e Silva, 1 and 2], [da Silva Dias], [Grothendieck], [Köthe, 1], and [Tillmann], all in the 1950's. See also [Gauthier and Rubel, 2].

Exercises

1. Prove that the seminorms $\|a\|_r$ and $\|a\|'_r$ generate the same topology on A.

2. Let E be a vector space and P a family of seminorms on E. Let $\tilde{P}$ be the family of seminorms described in the paragraph just prior to Proposition 6.1. Show that P and $\tilde{P}$ generate the same topology on E.

3. Find the dual space of $H(\mathbb{D} \times \mathbb{D})$.

4. Show that Definition 6.8 defines an equivalence relation.

5. Show that $H(\{0\})$ is in one-to-one correspondence with the collection of sequences $\{a_n\}$ of complex numbers satisfying $\limsup |a_n|^{1/n} < +\infty$.

6. Represent the dual of $H(\mathbb{C})$ as a space of germs of holomorphic functions.

7. For each of the following linear functionals on $H(\mathbb{D})$, represent L as a λ and as an L_F

   a) $L(f) = f(0)$

   b) $L(f) = \frac{1}{2\pi} \int_{|z|=\frac{1}{2}} \frac{f(z)}{z} dz$

   c) $L(f) = f'(0)$

   d) $L(f) = \int_0^{\frac{1}{2}} f(x)dx$

   e) $L(f) = \int_0^1 f(x)dx$

   f) $L(f) = \int_{|z|=\frac{1}{2}} \frac{f(z)}{z^2 + a} dz$

   g) $L(f) = \sum_n f^{(n)}(0)/(n!)^2$

8. Show that the operations $[f] + [g] \equiv [f + g]$, $[f][g] \equiv [fg]$, and $[f]' \equiv [f']$ are well-defined on $H(A)$.

9. Prove the uniqueness theorem for functions in $H(A)$. That is, suppose $\{z_n\}$ is a sequence of distinct points in A that converges to a point in A. Show that if $[f] \in H(A)$ satisfies $f(z_n) = 0$ for each n, then $[f] = 0$.

10. Prove that if $L, L' \in H(\mathbb{D})^*$ and if $L(z^n) = L'(z^n)$ for $n = 0,1,2,\ldots,$ then $L = L'$. Suppose only for $n = 1,2,3,\ldots;$ what can be said about the relation between L and L'?

11. Let $L \in H(\mathbb{D})^*$ and define $L'(f) = L(f')$. Given representations of L as a sequence $\{\lambda_n\}$ and as a germ F, what are the corresponding representations of L'?

§7. The Hahn-Banach Theorem, and Applications

A major tool in the application of duality results (in any locally convex topological vector space) is the Hahn-Banach Theorem. We state here one standard version (there are many equivalent versions) and two important corollaries.

(7.1) Theorem (Hahn-Banach). If E is a locally convex topological vector space, A is a subspace of E, and $L : A \to \mathbb{C}$ is a continuous linear functional on A, then L has an extension $\tilde{L}$ in E^*. i.e. $\tilde{L}$ is a continuous linear functional on E and $\tilde{L}(a) = L(a)$ for all $a \in A$.

(7.2) Corollary: If E is a locally convex topological vector space, A is a closed subspace of E, and $b \in E$, $b \notin A$, then there is an $L \in E^*$ such that $L|_A = 0$ and $L(b) = 1$.

(7.3) Corollary: If E is as above and A is a subspace of E, then A is dense in E if and only if the only functional $L \in E^*$ that vanishes on A is the zero functional (which vanishes on all of E).

Proof of the Corollaries: Given A and b as in the first corollary define a linear functional L_0 on $B = \{a + \gamma b : a \in A, \gamma \in \mathbb{C}\}$ by $L_0(a + \gamma b) = \gamma$. It is an exercise to prove that L_0 is continuous. By the Hahn-Banach Theorem there is an $L \in E^*$ that extends L_0.

One half of the second corollary is trivial: if $A^- = E$, $L \in E^*$ and $L|_A = 0$ then $L \equiv 0$ by continuity. For the other direction, if $A^- \neq E$ then the first corollary produces a functional in E^* that is not zero but vanishes on A^-. QED

We will be content with a sketch of a proof of the Hahn-Banach Theorem. The reader desiring more detail and other versions may consult almost any text on functional analysis.

Proof: (Sketch) The topology on E is generated by a family of seminorms and the topology on A is generated by the restrictions of those seminorms to A. It thus suffices to prove that if L is a functional on A and $\|\cdot\|$ is a seminorm on E with $|L(a)| \leq \|a\|$ for all $a \in A$, then there is an extension $\tilde{L} \in E^*$ such that $|\tilde{L}(f)| \leq \|f\|$ for all $f \in E$.

Let $b \notin A$ and let B be as in the proof of the first corollary.

Extend L to B by defining $\tilde{L}(a + \gamma b) = L(a) + \gamma\beta$ where β must be chosen to preserve the inequality on L. We first extend $\mathrm{Re}(L)$ to a functional that is linear for real scalars: We wish to choose a real number β such that

$$-\|a + \gamma b\| \le \mathrm{Re}(L(a)) + \gamma\beta \le \|a + \gamma b\|$$

for all $a \in A$, $\gamma \in \mathbb{R}$. The assumed inequality on A implies that

$$\mathrm{Re}\, L(x + y) \le \|x + b - (b - y)\|$$

so

$$\sup_{y\in A}(\mathrm{Re}\, L(y) - \|b - y\|) \le \inf_{x\in A}(\|x + b\| - \mathrm{Re}\, L(x)).$$

Pick any β between these two extremes. Repeat the procedure with ib to get $\mathrm{Re}(L)$ defined on B. To define $\tilde{L}$ it suffices to put

$$\tilde{L}(a + \gamma b) = \mathrm{Re}\, L(a + \gamma b) - i\, \mathrm{Re}\, L(ia + i\gamma b)$$

for $\gamma \in \mathbb{C}$.
This extends L to a larger subspace. An application of the Hausdorff Maximality Principle now shows that L has a maximal extension, which must be defined on all of E. QED.

(7.4) Let us now sketch a trivial application of the Hahn-Banach Theorem to holomorphic functions. We use it to prove that the polynomials are dense in $H(\mathbb{D})$. (It should be remembered that we have actually used a stronger fact than this to obtain the characterization of $H(\mathbb{D})^*$. Our only reason for including this example is to illustrate a general method of proof.) Using the second corollary to the Hahn-Banach Theorem, it suffices to show that any $L \in H(\mathbb{D})^*$ which vanishes on the subspace of polynomials, vanishes everywhere on $H(\mathbb{D})$. But, if $L(z^n) = 0$ for $n = 0,1,2,\ldots$ then the correspondence between $H(\mathbb{D})^*$ and $H_0(\mathbb{C}\hat{}\setminus\mathbb{D})$ gives $0 = L(z^n) = \frac{1}{2\pi i}\int_\Gamma F(z)z^n dz = F^{(n+1)}(\infty)/(n + 1)!$ Thus F and all of its derivatives vanish at ∞. Since $\mathbb{C}\hat{}\setminus\mathbb{D}$ is connected, F must be 0, hence $L = 0$.

We get a bonus here because this proof goes through, word for word, in case we replace $\mathbb{D}$ by any finite union of disjoint disks, provided we replace $\int_\Gamma$ by $\sum_j \int_{\Gamma_j}$ (Theorem 7.5, below). In fact this method is quite general. We will use it in Chapter 10 to prove that if G is any open set in $\mathbb{C}$ such that $\mathbb{C}^\wedge \setminus G$ is connected, then $P^- = H(G)$, where P denotes the polynomials. The proof is the same once we have shown that $H(G)^* = H_0(\mathbb{C}^\wedge \setminus G)$ in the appropriate sense.

Let us formally state the result on disjoint disks.

(7.5) Theorem: The polynomials are dense in $H(G)$ if G is a finite union of disjoint disks.

Proof. This is left to the reader. The proof is almost word for word the same as (7.4). The only part that might not be obvious is showing that if K is a closed connected set, $[F]$ is a germ on K and F together with all its derivatives vanishes at a point of K, then $[F] = 0$. This is proved as Proposition 10.1 and is left as an exercise for now.

To illustrate this result let

$$D_1^- = \{z : |z| \leq 1\}$$

$$D_2^- = \{z : |z - 5| \leq 2\}$$

$$D_3^- = \{z : |z - 3i| \leq 1/2\}.$$

Then there is a single polynomial $P(z)$ such that

$$|P(z)| < 10^{-10} \quad \text{for} \quad z \in D_1^-,$$

$$|P(z) - e^z| < 10^{-10} \quad \text{for} \quad z \in D_2^-, \quad \text{and}$$

$$\left|P(z) - \sin\left(\frac{z^2 + 1}{z^2 - 1}\right)\right| < 10^{-10} \quad \text{for} \quad z \in D_3^-.$$

Our proof does have the drawback of being non-constructive so that we have no idea of what P looks like. This result is a special case of Runge's Theorem, which we prove in a later chapter. A constructive proof of Runge's Theorem is quite well known and is the standard proof in many texts. See the Notes to Chapter 10.

As a consequence of the preceding results we prove:

(7.6) Proposition. There is a sequence $\{f_n\}$ of entire functions that converges at each point in $\mathbb{C}$ to 0, but does not converge uniformly on compact sets.

(7.7) <u>Proposition</u>: <u>There is a sequence</u> $\{g_n\}$ <u>of entire functions that converges pointwise in</u> $\mathbb{C}$ <u>to a discontinuous function</u>.

<u>Proofs</u>: (It will help to draw pictures of the following sets.) Let

$$A_n = \{z : |z + n!| < n! + \frac{2}{n}\}$$

$$B_n = \{z : |z - \frac{4}{n}| < \frac{1}{n}\}$$

$$C_n = \{z : |z - (n! + \frac{7}{n})| < n! + \frac{1}{n}\}$$

$$K_n = \{z : |z + n!| \le n! + \frac{1}{n}\} \subseteq A_n$$

$$L_n = \{z : z = \frac{4}{n}\} \subseteq B_n$$

$$M_n = \{z : |z - (n! + \frac{7}{n})| \le n!\} \subseteq C_n.$$

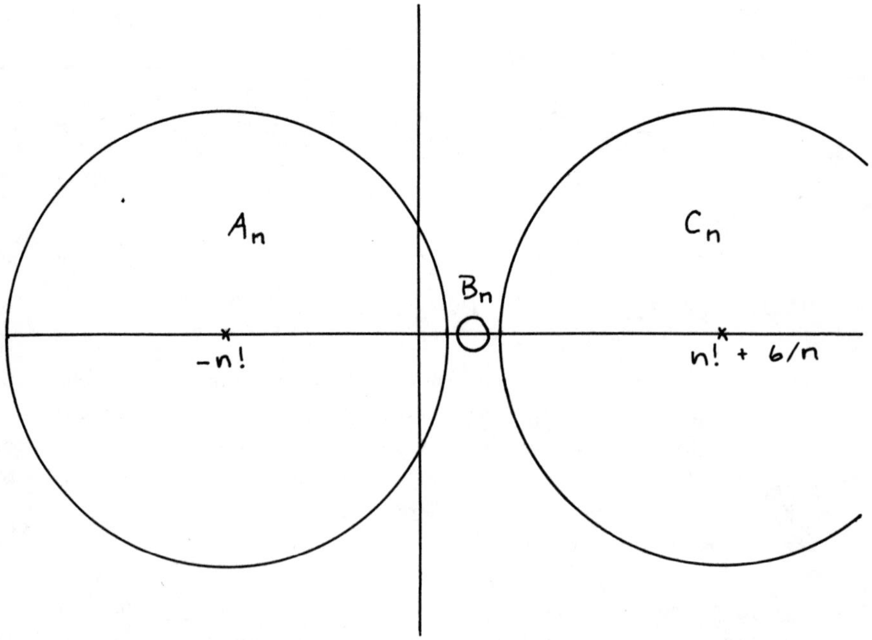

Fig. 7.1.

Then A_n, B_n, and C_n are disjoint open disks and K_n, L_n, and M_n are compact subsets. Define $f_n \in H(A_n \cup B_n \cup C_n)$ by

$$f_n(z) = \begin{cases} 0 & z \in A_n \\ 1 & z \in B_n \\ 0 & z \in C_n \end{cases}.$$

Choose polynomials P_n such that $|P_n(z) - f_n(z)| < \frac{1}{n}$ for all $z \in K_n \cup L_n \cup M_n$. Since each point $z \in \mathbb{C}$ is eventually in $A_n \cup C_n$ (why?) it is evident that $P_n(z) \to 0$ at each point z but not uniformly on any compact set that contains $\{\frac{4}{n} : n = 1,2,\ldots\}$. For the second proposition, let

$$g_n(z) = \begin{cases} 0 & z \in A_n \\ \\ 1 & z \in C_n \end{cases}.$$

Then $g_n \in H(A_n \cup C_n)$ and so polynomials Q_n exist with $|Q_n(z) - g_n(z)| < \frac{1}{n}$ for all $z \in K_n \cup M_n$. Clearly $g_n(z) \to g(z)$ where

$$g(z) = \begin{cases} 0 & \operatorname{Re} z \leq 0 \\ \\ 1 & \operatorname{Re} z > 0 \end{cases}$$

and so $Q_n(z)$ converges pointwise to the discontinuous function $g(z)$. QED

It would seem that holomorphic functions can converge pointwise to almost anything. However, things are not quite as bad as that, since we have the following theorem of Osgood.

(7.8) Theorem (Osgood). *Given an open set G in $\mathbb{C}$ and a sequence $\{f_n\}$ of functions in $H(G)$ that converges pointwise to f, then there is a dense open set $G' \subseteq G$ such that $f_n|_{G'}$ converges to $f|_{G'}$ in $H(G')$. In particular f must be holomorphic in a dense open subset of G.*

Proof: Let $G_n = \{z : |f_k(z)| \leq n$ for $k = 1,2,3,\ldots\}$. Then G_n is a closed subset of G and $G = \cup G_n$. By the Baire Category Theorem some G_n is not nowhere dense, i.e. contains an open disk Δ. Let G'

be the union of all open disks Δ contained in some G_n (in fact $G' = \cup \operatorname{int} G_n$). Then G' is open and dense (otherwise apply the same reasoning to $G\setminus (G')^-$). The family $\{f_n|_{G'}, n = 1,2,3,\ldots\}$ is a bounded family in $H(G')$ (Why?). The proof is then finished by the following lemma.

(7.9) Lemma. If $\{f_n\}$ is a sequence bounded in $H(G)$ where G is open, and if $\{f_n\}$ converges pointwise to f, then $\{f_n\}$ converges to f in $H(G)$.

Proof: First of all, $\{f_n\}$ must have a convergent subsequence in $H(G)$. Its limit must agree with the pointwise limit, so $f \in H(G)$. Suppose $f_n - f$ does not converge to zero in $H(G)$. Then there exist a compact set $K \subset G$, and $\varepsilon > 0$, and a subsequence $\{f_{n_k}\}$ such that $\|f_{n_k} - f\|_K > \varepsilon$ for all k. But $\{f_{n_k}\}$, being bounded in $H(G)$, has a convergent subsequence and its limit must again be f. Thus $f_{n_k} - f$ has a subsequence converging to 0 in $H(G)$. This contradicts the requirement that $\|f_{n_k} - f\|_K > \varepsilon$ for all k, and so $f_n - f \to 0$ in $H(G)$. QED

(7.9) Non-continuable functions.

Given $f \in H(G)$, we say f is non-continuable if, given any $z \in \partial G$ and any disk D centered at z, f does not have a holomorphic extension to $G \cup D$. We also say that ∂G is a natural boundary for f in this case. We have seen earlier that, in the case of two variables when G is a polydisk with center removed, there are no non-continuable functions in $H(G)$.

(7.10) Proposition. Let $f \in H(\mathbb{D})$ be defined by $f(z) = \sum_{n=1}^{\infty} z^{n!}$. Then f is non-continuable.

Proof: We show that $\lim_{r\to 1} f(re^{i\theta}) = \infty$ for each θ of the form $\theta = 2\pi\rho$ where $\rho = p/q$ is a real rational number. For then $f(re^{i\theta}) = \sum r^{n!} \exp(2\pi i \frac{p}{q} n!)$ and $n!\, p/q$ is an integer for $n \geq q$. Thus $f(re^{i\theta}) = \sum_{n<q} r^{n!} \exp(2\pi i\rho n!) + \sum_{n\geq q} r^{n!}$ and the second sum tends to $+\infty$ as $r \to 1$. This proves the proposition because any open proper superset of $\mathbb{D}$ must contain a point where f tends to ∞. QED

(7.11) Proposition. There is a non-continuable function $f \in H(\mathbb{D})$ that has a continuous extension to the whole plane.

Proof: Let $f(z) = \sum_{n=1}^{\infty} (1 + n!)^{-1} z^{n!+1}$. This series converges uniformly in $\mathbb{D}^-$ and so f is continuous there. It can then be extended to the whole plane by the Tietze Extension Theorem (2.27). It is non-continuable because its derivative $\sum_{n=1}^{\infty} z^{n!}$ tends radially to ∞ on a dense subset of $\partial\mathbb{D}$. QED

(7.12) Theorem. In $H(\mathbb{D})$, the complement of the set of non-continuable functions is a set of first category.

In a sense, this says "most" functions in $H(\mathbb{D})$ are non-continuable. Since $H(\mathbb{D})$ is a Baire space (second category in itself) the set of non-continuable functions is non-empty and contains the intersection of a countable family of dense open sets. This theorem therefore proves that non-continuable functions exist without actually exhibiting one.

Proof. Given $0 < \varepsilon < 1$ and $0 < \rho < 1$ let

$$A_\rho(\varepsilon) = \{f \in H(\mathbb{D}) : \forall\, \theta\ \exists\, r > \rho \text{ such that } |f(re^{i\theta})| < \varepsilon\}$$

and

$$A'_\rho(\varepsilon) = \{f \in H(\mathbb{D}) : \forall\, \theta\ \exists\, r > \rho \text{ such that } |f(re^{i\theta})| > \tfrac{1}{\varepsilon}\}.$$

The goal is to prove that each $A_\rho(\varepsilon)$ and $A'_\rho(\varepsilon)$ is open and dense in $H(\mathbb{D})$. The intersection of these sets for a sequence of ρ's tending to one and some ε less than one consists entirely of functions f such that $\lim_{r\to 1^-} f(re^{i\theta})$ does not exist for any θ. Such functions are non-continuable. The complement of this intersection is a set of first category.

First, each $A_\rho(\varepsilon)$ is open. Given $f \in A_\rho(\varepsilon)$, let $J = \{z \in \mathbb{D} : |f(z)| < \varepsilon\}$. J is open and intersects every radius at some point outside $D_\rho = \{z : |z| < \rho\}$. Thus, for each real θ, there is an open disk Δ', with the center of Δ' being $re^{i\theta}$, $\rho < r < 1$, and $\Delta' \subset \mathbb{D}\setminus D_\rho$ and $\Delta' \subset J$. Let Δ be the disk with the same center as Δ' but with half the radius. Let $A = \{z/|z| : z \in \Delta\}$, A is open and, as θ varies, the totality of A's cover ∂D. By compactness there exist finitely many Δ's, $\Delta_1, \Delta_2, \ldots, \Delta_n$, such that the corresponding $A_1, \ldots, A_n$ cover $\partial\mathbb{D}$. Let $r' = \max\{|z| : z \in \cup\bar{\Delta}_j\}$, so that $r' < 1$ and $r' > \rho$. By the maximum modulus principle there exists an $\varepsilon' > 0$, $\varepsilon' < \varepsilon$ such that $|f(z)| < \varepsilon'$ for $z \in \cup\Delta_j$. Let K be a compact set containing $\cup\Delta_j$. If $\|f - g\|_K < \frac{\varepsilon - \varepsilon'}{2}$, then $g \in A_\rho(\varepsilon)$. Thus $A_\rho(\varepsilon)$ is open.

Next, we show $A_\rho(\varepsilon)$ is dense. Given any $f \in H(\mathbb{D})$ and $\delta < 0$ (we may suppose $\delta < \varepsilon$) and any compact $K \subset \mathbb{D}$ we must find $g \in A_\rho(\varepsilon)$ with $\|g - f\|_K < \delta$. Choose $R < 1$ with $R > \rho$ and $K \subset D_R = \{z : |z| < R\}$. Let $\mathcal{D}$ be a finite union of disjoint disks Δ_j such that the disks Δ_j' (with the same centers and twice the radius) are disjoint and such that a) each radius of $\mathbb{D}$ intersects $\mathcal{D}$ and b) $\Delta_j' \subset \mathbb{D}\setminus D_R$ for all j. Let $G = D_R \cup \bigcup_j \Delta_j'$. Define $g \in H(G)$ by $g(z) = f(z)$ when $z \in D_R$ and $g(z) = 0$, $z \in \Delta_j'$, all j. By Runge's theorem, there is a polynomial p such that $|p(z) - g(z)| < \delta$ for $z \in K \cup \mathcal{D}$. It follows that $p \in A_\rho(\varepsilon)$ and $\|p - f\|_K < \delta$.

Thus, each $A_\rho(\varepsilon)$ is dense and open. Similarly $A_\rho'(\varepsilon)$ is dense and open. Let $\rho_n \to 1$ and $\varepsilon < 1$. The sets $A_{\rho_n}(\varepsilon) \cap A_{\rho_n}'(\varepsilon)$ are then countably many dense open sets whose intersection consists of functions which on every radius take on both values less than ε and values greater than $1/\varepsilon$. Such functions do not have even a continuous extension to any point of $\partial\mathbb{D}$ and so are non-continuable. QED

(7.13) Remark: A slight modification of the above prodecure shows that for "most" functions $f \in H(\mathbb{D})$, the curve $f(re^{i\theta})$, $0 \le r < 1$ is dense in the plane for each θ. It is also true that most functions map each sector onto the whole complex plane.

We will see later (an exercise following Chapter 12) that for each open $G \subseteq \mathbb{C}$ there is a non-continuable function $f \in H(G)$. For several variables the situation is more complicated.

If a power series $f(z) = \sum_{n=0}^{\infty} a_n z^n$ has the unit circle as its circle of convergence, and if "many" of the coefficients a_n are zero, then f is non-continuable. The strongest theorem to this effect is the next result, which we present without proof.

(7.14) Fabry Gap Theorem. *Suppose $f(z) = \sum a_k z^{n_k}$ has the unit circle as its circle of convergence, and suppose that $n_k/k \to \infty$ as $k \to \infty$. Then f is a non-continuable function in $H(\mathbb{D})$.*

NOTES: A proof of the Fabry Gap Theorem can be found in [Hille, Vol.

II, Theorem 11.7.2, p. 89ff].

More detailed proofs of and more versions of the Hahn-Banach Theorem can be found in any text on functional analysis, including [Köthe, 2] and [Kelley and Namioka].

Exercises

1. Let A be a closed subspace of the locally convex space E and $b \notin A$. Show that the functional $L(a + \gamma b) = \gamma$ defined on $\{a + \gamma b : a \in A, \gamma \in \mathbb{C}\}$ is continuous. (Hint: It needs to be shown that for any $\varepsilon > 0$ there is a nbd V of 0 in E such that $|\gamma| < \varepsilon$ whenever $a + \gamma b \in V$. If not show that $\exists \varepsilon > 0$ such that $(b + \frac{1}{\varepsilon} V) \cap A \neq \emptyset$ for all nbds V of 0 in E and so $b \in A^- = A$.)

2. Show that there exists a sequence of polynomials $\{p_n\}$ such that

$$\lim_{n\to\infty} p_n(x) = \begin{cases} 1 & \text{if } \exists \text{ integer } k \text{ with } 2k < x < 2k + 1 \\ -1 & \text{if } \exists\, k \text{ with } 2k - 1 \leq x \leq 2k \end{cases}$$

3. Find a continuous linear functional L on $C(\mathbb{D})$ which extends $L(f) = f'(0)$ defined on $H(\mathbb{D})$.

4. Show that the polynomials are dense in $H(G)$ where G is the half-plane $\operatorname{Re} z > 0$. (Use the corresponding result for disks.)

5. Suppose $L \in H(\mathbb{D})^*$, $L \neq 0$. Must $L(f) \in f(\mathbb{D})$?

6. Prove that for any sequence $\{A_n\}$ of positive real numbers satisfying $\limsup A_n^{1/n} \geq 1$, there exists a non-continuable function $f(z) = \sum a_n z^n$ in $H(\mathbb{D})$ such that $|a_n| \leq A_n$ for every n.

§8. More Applications

In this section E will denote the space $H(\mathbb{C})$, i.e., the space of entire functions with the topology of uniform convergence on compact sets.

(8.1) Definition: If $f \in E$ and $\tau \in \mathbb{C}$, we define $f_\tau \in E$ by $f_\tau(z) = f(z - \tau)$ and we call f_τ the translate of f by τ.

For example, $\sin z$ is the translate of $\cos z$ by $\pi/2$: $\cos(z - \pi/2) = \sin z$.

(8.2) Proposition: (Birkhoff) There is an entire function f whose translates are dense in E.

We might call such a function "universal". Roughly speaking, given any compact piece of the graph of any entire function g whatsoever, we can cut a piece out of the graph of f that nearly matches that of g.

Proof: Let $D^{(n)}$, $n = 1,2,3,\ldots,$ be a sequence of disks whose closures are disjoint and whose radii approach ∞ with n. We suppose also that there are disks U_n, $n = 1,2,3,\ldots,$ centered at 0, such that U_n contains the closures of $D^{(1)},\ldots,D^{(n)}$ but $\bar{U}_n$ is disjoint from the closures of $D^{(n+k)}$ for $k \geq 1$. For instance let

$$D^{(n)} = \{z : |z - (n+1)^2| < n - 1\}$$

$$U_n = \{z : |z| < (n+1)^2 + n\}$$

Let $\{p_n\}_{n=1}^{\infty}$ be an enumeration of the polynomials with rational coefficients. Then $\{p_n : n = 1,2,\ldots\}$ is dense in E. If γ_n is the center of $D^{(n)}$, let q_n be the translate of p_n by γ_n, that is, $q_n = (p_n)_{\gamma_n}$. Let $f_1 = q_1$ and define the polynomials f_n inductively so that

$$|f_n(z) - f_{n-1}(z)| < 2^{-n} \quad \text{for} \quad z \in U_n$$

and

$$|f_n(z) - q_n(z)| < 2^{-n} \quad \text{for} \quad z \in D^{(n)}.$$

Such polynomials exist because of our special case of Runge's Theorem for disjoint disks (Theorem 7.5). Notice that, for $m > n$

$$|f_m(z) - f_n(z)| \le |f_m(z) - f_{m-1}(z)| + \cdots + |f_{n+1}(z) - f_n(z)|$$

$$\le 2^{-m} + \cdots + 2^{-n-1}$$

$$\le 2^{-n}$$

for all $z \in U_n$. By the Cauchy criterion f_n converges in E to some $f \in E$. It is an exercise for the reader to show that f is universal. QED

Our next application is a result on approximation by exponentials.

(8.3) Definition: In a vector space V, if F is a subset of V then the span of F, written span F, is the set of all finite linear combinations of elements of F, that is, all finite sums $\sum a_j f_j$, $a_j \in \mathbb{C}$ and $f_j \in F$. The closed span of F is the closure of span F.

(8.4) Theorem: Given a sequence of distinct complex numbers $\{\lambda_n\}$, each $\lambda_n \neq 0$, let $\lambda(t) = \sum_{|\lambda_n| \le t} 1$ be the number of λ_n with $|\lambda_n| \le t$. If $\limsup_{r\to\infty} \frac{1}{r} \int_0^r \frac{\lambda(t)}{t}\,dt \ge 1$, then the closed span of $\{e^{\lambda_n z} : n = 1,2,3,\ldots\}$ is all of $H(\mathbb{D})$ ($\mathbb{D}$ being the unit disk).

Our proof of this theorem requires the following classical result.

(8.5) Jensen's Theorem: If f is holomorphic on $|z| \le r$ and if the Taylor expansion of f around 0 is given by $f(z) = a_k z^k + a_{k+1} z^{k+1} + \cdots$, $a_k \neq 0$, then

$$\frac{1}{2\pi} \int_{-\pi}^{\pi} \log |f(re^{i\theta})|\,d\theta = \log|a_k| + k \log r + \sum_{r_n \le r} \log \frac{r}{r_n}$$

where the zeros of f are $z_j = r_j e^{i\theta_j}$, repeated according to their multiplicity, and not counting the zero at 0.

Proof: Without loss of generality, take $r = 1$. Also, suppose that f has no zeros on $|z| = 1$. (The general case follows by continuity of both sides of the equality as functions of r.) Let

$$g(z) = z^k \prod_{r_n \le 1} \frac{z - z_n}{1 - \bar{z}_n z}, \quad \text{so that} \quad h = f/g$$

is holomorphic on $|z| \le 1$ and has no zeros there. Consequently, there is a holomorphic branch of $\log h$ on $|z| \le 1$, that is, there is a function H holomorphic on $|z| \le 1$ such that $h = \exp H$. (Exercise 15, Chapter 2.) Now,

$$H(0) = \frac{1}{2\pi i} \int_{|z|=1} \frac{H(z)dz}{z} = \frac{1}{2\pi} \int_{-\pi}^{\pi} H(e^{i\theta})d\theta$$

so that

$$\mathrm{Re}H(0) = \frac{1}{2\pi} \int_{-\pi}^{\pi} \mathrm{Re}\; H(e^{i\theta})d\theta.$$

But

$$\mathrm{Re}H = \log|h| = \log\left|\frac{f}{g}\right|.$$

Hence

$$\log(|a_k|/\prod_{r_n \le r} |z_n|) = \frac{1}{2\pi} \int_{-\pi}^{\pi} \log|f(e^{i\theta})|d\theta - \frac{1}{2\pi} \int_{-\pi}^{\pi} \log|g(e^{i\theta})|d\theta.$$

But, as $|g(z)| = 1$ when $|z| = 1$, the second integral is zero, and the result is proved. QED

Remark: If $n(t)$ = number of zeros of f in $|z| \le t$, then

$$\sum_{r_n \le r} \log \frac{r}{r_n} = \sum \int_{r_n}^{r} \frac{dt}{t} = \int_0^r \frac{n(t) - n(0)}{t}\, dt$$

Proof: (Of Theorem 8.4.)

By the Hahn-Banach Theorem, it is enough to prove that if $L \in H(\mathbb{D})^*$ and $L(\exp(\lambda_n z)) = 0$ for $n = 1,2,3,\ldots$, then $L = 0$. We first prove that $L(e^{wz}) = 0$ for all w. To this end, let $L^\wedge(w) = L(e^{wz})$. Then $L^\wedge$ is an entire function and satisfies the inequality

(*) $|L^\wedge(w)| < Ce^{\rho|w|}$ for some $\rho < 1$,

where C is a positive constant. To see this, let $w \in \mathbb{C}$ be arbitrary and compute

$$(**)\quad \frac{L^\wedge(w') - L^\wedge(w)}{w' - w} = L\left(\frac{e^{w'z} - e^{wz}}{w' - w}\right),$$

and observe that $\frac{e^{w'z} - e^{wz}}{w' - w} \to ze^{wz}$ in $H(\mathbb{D})$ as $w' \to w$. This means the difference quotient in (**) tends to $L(ze^{wz})$, i.e. $L^\wedge$ is entire. To prove the inequality (*), we use that fact that there is a compact set $K \subseteq \mathbb{D}$ and a constant C such that $|L(f)| \leq C\|f\|_K$ for all $f \in H(\mathbb{D})$. This is simply the continuity condition for L. We may suppose that $K \subseteq \{z : |z| \leq \rho\}$ for some $\rho < 1$ and then

$$|L^\wedge(w)| = |L(e^{wz})| \leq C \sup\{|e^{wz}| : |z| \leq \rho\}$$

$$\leq C\, e^{\rho|w|}.$$

Now we apply Jensen's Theorem and use the fact that each λ_n is a zero of $L^\wedge$. This gives

$$\int_0^r \frac{\lambda(t)}{t}\,dt \leq \text{const.} + \rho r + \text{const.} \log r,$$

which implies $\limsup_{r\to\infty} \frac{1}{r}\int_0^r \frac{\lambda(t)}{t}\,dt \leq \rho < 1$, contrary to assumption. The only way this contradiction can be resolved is to conclude $L^\wedge$ is an entire function to which Jensen's Theorem is inapplicable, i.e. $L^\wedge = 0$. By differentiating we find that

$$0 = \frac{d^n}{dw^n} L^\wedge(w) = L(z^n e^{zw}).$$

Setting $w = 0$ gives $L(z^n) = 0$, $n = 1,2,\ldots$ and it follows that $L = 0$. QED

(8.6) <u>Remarks</u>. The reader might benefit by using the sequence interpretation of the dual of $H(\mathbb{D})$: If L is associated with the sequence $\{a_k\}$, $\limsup|a_k|^{1/k} < 1$. Then

$$L^\wedge(w) = L(e^{wz}) = L(\textstyle\sum(1/k!)w^k z^k)$$

$$= \sum \frac{1}{k!}\, a_k w^k$$

It is then easy to see that $L^\wedge$ is entire and the estimate $|L^\wedge(w)| \leq \sum(1/k!)|a_k|\;|w|^k \leq C\exp[|w|(\limsup|a_k|^{1/k} + \varepsilon)]$ is not hard.

NOTES: Our construction of a "universal" entire function (Proposition 8.2) appears in [Seidel and Walsh].

The process of forming the "Laplace transform" of a linear functional ($L^{\wedge}(w)$ in the proof of Theorem 8.4) is quite useful. Its usefulness depends mostly on having information about the space of entire functions of exponential type (inequality (*) in the same proof), especially about their zeros.

Jensen's Theorem is another useful tool, primarily for furnishing information about the behavior of zeros of analytic functions in a disk, when a restriction on the "size" of the function is imposed.

Exercises

1. Show that the function f obtained in the proof of Proposition 8.2 is universal.

2. Verify that the span of $\{e^{nz} : n = 1,2,3,\ldots\}$ is dense in $H(\mathbb{D})$.

3. Is the span of $\{e^{nz} : n = 1,2,3,\ldots\}$ dense in $H(G)$ when $G = \{z : |z| < 10\}$? (Consider the periodicity of $\exp(nz)$.)

4. Suppose f is an entire function such that there exist constants C, ρ with $0 < \rho < 1$ and

$$|f(z)| < C\, e^{\rho|z|}, \quad \text{all } z \in \mathbb{C}.$$

Prove that $f(z) = L^{\wedge}(z)$ for some $L \in H(\mathbb{D})^*$. (Hints: Write $f(z) = \sum_{n=0}^{\infty} \frac{a_n}{n!} z^n$ and estimate $|a_n|$ by Cauchy's inequalities. Minimize over r to obtain $|a_n| \le \rho^n C\, e^n n!/n^n$. Obtain $\lim |a_n|^{1/n} \le \rho$--using Stirlings formula, perhaps. Now associate a linear functional L with $\{a_n\}^{\infty}$, and show $f(w) = L(e^{wz})$.)

5. Identify E with the space of sequences $\{\{a_n\}_{n=0}^{\infty} : \limsup |a_n|^{1/n} = 0\}$. Show that the dual E^* may be identified with the space of sequences $\{\{b_n\}_{n=0}^{\infty} : \limsup |b_n|^{1/n} < +\infty\}$.

(Hint: The action of $\{b_n\}$ on $\{a_n\}$ is to be $\sum a_n b_n$. If $L \in E^*$, let $b_n = L(z^n)$. Show that $\limsup |b_n|^{1/n} < +\infty$. Conversely, if the lim sup is finite show $\sum a_n b_n$ converges and defines a continuous functional on E.)

6. Show that there exists a function $f \in E$ such that $\operatorname{span}\{f(nz) : n = 0,1,2,\ldots\}$ is dense in E.

Outline: Need to choose $\{a_k\}_0^\infty$ with $\limsup |a_k|^{1/k} = 0$ such that if $L(\sum_k a_k n^k z^k) = 0$ for all n, then $L = 0$. Associate L with $\{b_k\}$ as in Exercise 5 and let $g(z) = \sum a_k b_k z^k$. Then $L(\sum a_k n^k z^k) = g(n)$.

Consequently, it suffices to choose $\{a_k\}$ so that if $\{b_k\}$ satisfies $\limsup |b_k|^{1/k} < +\infty$ and $g(n) = 0$ for $n = 0,1,2,\ldots,$ then $g \equiv 0$.

Choose $a_k = (k^k k!)^{-1}$. Then if $\limsup |b_k|^{1/k} < R$, it follows that

$$\begin{aligned} \left|\sum a_k b_k z^k\right| &\le C_1 \sum \frac{(R/k)^k |z|^k}{k!} \\ &\le C_\varepsilon e^{\varepsilon |z|}, \quad \varepsilon < 1 \end{aligned}$$

It then follows, as in the proof of 8.4, that if g is not identically zero and if $n(t)$ is the number of zeros of g with modulus less than or equal to t, then $n(t) \ge [t]$ so

$$\begin{aligned} 1 &= \limsup_{r\to\infty} \frac{1}{r} \int_0^r \frac{[t]}{t}\, dt \\ &\le \limsup_{r\to\infty} \frac{1}{r} \int_0^r \frac{n(t)}{t}\, dt \le \varepsilon, \end{aligned}$$

a contradiction.

7. Show that E^* can be identified with the functions of exponential type. (An entire function f is said to be of <u>exponential type</u> if there are constants A and B such that

$$|f(z)| \le A e^{B|z|}, \quad \text{all } z \in \mathbb{C}.)$$

(Compare with Exercises 4 and 5.)

*8. Does there exist $L \in E^*$ such that $L \ne 0$ and

a) $L(f) = L(f')$ for every $f \in E$?

b) for some f, $L(f^n) = L(f^{(n)})$, $n = 1,2,3,\ldots$?

9. Given $f, g, h \in E$, suppose $L(f) = L(g)$ for every $L \in E^*$ such that $L(h) = 0$. What can you say about the relationship between f, g and h?

*10. Suppose given $f \in E$, f not a constant and $L \in E^*$ and suppose

$L(f^n) = [L(f)]^n$ for $n = 1,2,3,\ldots$. What can be said about L and f? (Given f and points $z_1, z_2, \ldots, z_k$ such that $f(z_i) = f(z_j)$ for all $i,j \leq k$, an example of such an L is $L(g) = \sum a_i g(z_i)$ where $\sum a_i = 1$. Must L always have this or a similar form?)

11. Is there an $L \in H(\mathbb{D})^*$ so that

a) $L(z^n) = n$?

b) $L(z^n) = \frac{1}{n}$?

c) $L(z^n) = n!$?

12. Same question for E^*.

§9. The Dual of $H(G)$

(9.1) We want to prove, as in the case of the disk, that $H(G)^* = H_0(\hat{\mathbb{C}}\setminus G)$. We first study the dual of $C(G)$. We change our notation here and write $L(f) = \int f d\mu$ when $L \in C(G)^*$. (For the reader unfamiliar with integration theory this is simply a change in notation: The left-hand side defines the right-hand side. There are two advantages to this notation. First, it is the notation in which research papers are written. Second, the reader can call upon her experience with integration for intuition. For the mathematically advanced reader: we are invoking the Riesz Representation Theorem for $C(G)^*$.) We call μ the "measure" associated with L, and we may identify μ and L. The collection of all such μ is denoted $M_0(G)$, so that $M_0(G) = C(G)^*$. We also write $L(f) = \int f(z)d\mu(z)$ when it is necessary to indicate the independent variable. "Measures" have the same properties as continuous linear functionals (which is what they are); for reinforcement, we list them here. Given $\mu \in M_0(G)$:

i) $\int (f + g)d\mu = \int fd\mu + \int gd\mu$, $f, g \in C(G)$.

ii) $\int afd\mu = a \int fd\mu$, $f \in C(G)$, $a \in \mathbb{C}$.

iii) If $f_n \to f$ in $C(G)$ then $\int f_n d\mu \to \int fd\mu$.

iv) There is a compact set $K \subseteq G$ such that

$$\left|\int fd\mu\right| \le C\|f\|_K \quad \text{for all } f \in C(G).$$

(9.2) <u>Definition</u>: <u>If</u> K <u>is a compact set in</u> G <u>such that for some constant</u> C, $|\int fd\mu| \le C\|f\|_K$, <u>all</u> $f \in C(G)$, <u>we say that</u> K <u>supports</u> μ.

We digress here to point out that there exist continuous linear functionals on other spaces of continuous functions on G which are also called measures but do not satisfy the inequality in the definition. (This is the reason for the subscript in $M_0(G)$.) Thus we say that $M_0(G)$ consists of <u>measures with compact support</u>.

One consequence of this inequality is that $\int fd\mu$ is independent of the values of f on $G\setminus K$. That is, if $f, g \in C(G)$ and $f|_K = g|_K$, then $\int fd\mu = \int fd\mu$. To prove this, simply observe $|\int (f - g)d\mu| \le C\|f - g\|_K = 0$.

(9.3) <u>Definition</u>: <u>For</u> $\mu, \nu \in M_0(G)$ <u>we write</u> $\mu \sim \nu$ <u>to mean that</u> $\int fd\mu = \int fd\nu$ <u>for all</u> $f \in H(G)$.

Example: Take $G = \mathbb{D} = \{z : |z| < 1\}$, and let μ be the "point mass" at 0, i.e. $\int f d\mu = f(0)$ for all $f \in C(\mathbb{D})$. If $h \in H(\mathbb{D})$ and $\Gamma = \{z : |z| = \frac{1}{2}\}$ we have $h(0) = \frac{1}{2\pi i} \int_\Gamma h(z) \frac{dz}{z}$, so that $\mu \sim \nu$ where ν is the measure $\frac{1}{2\pi i} \frac{dz}{z}$ on Γ. It is clear that $\mu \neq \nu$ ($\int |z| d\mu \neq \int |z| d\nu$) and, in fact, μ is supported by $\{0\}$ while ν is supported by Γ. (It is easy to see that equal measures cannot have disjoint supports.)

It is clear that $\sim$ is an equivalence relation and so we write $[\mu]$ for the equivalence class containing μ. We use $M_0'(G)$ to denote the vector space of all equivalence classes $[\mu]$ for all $\mu \in M_0(G)$. As a preliminary (although not particularly deep) result we have the following.

(9.4) <u>Proposition</u>. $H(G)^* = M_0'(G)$. <u>This means there is a one-one correspondence between</u> $H(G)^*$ <u>and</u> $M_0'(G)$ <u>which associates to each</u> $[\mu] \in M_0'(G)$ <u>the functional</u> $L_{[\mu]}$ <u>defined by</u> $L_{[\mu]}(f) = \int f d\mu$.

<u>Proof</u>: Each such $L_{[\mu]}$ belongs to $H(G)^*$ because μ is continuous, by definition, on $C(G)$ and therefore on $H(G)$. Moreover, the definition of $L_{[\mu]}$ is independent of the measure chosen to represent $[\mu]$ because if $[\mu] = [\nu]$ then (by definition) $\int f d\mu = \int f d\nu$ for all $f \in H(G)$. To see that every $L \in H(G)^*$ arises in this way, use the Hahn-Banach Theorem to extend L to a continuous linear functional on $C(G)$ (and hence a measure $\mu \in M_0(G)$). Since μ extends L, we have $L(f) = \int f d\mu$ for $f \in H(G)$. Thus $L = L_{[\mu]}$.

The experienced reader will recognize this is nothing but $A^* = X^*/A^\perp$, where A is a closed subspace of X and $A^\perp$ denotes the linear functionals on X that vanish on A. The method of proof is quite general.

(9.5) <u>Main Duality Theorem</u>. $H(G)^* = H_0(\mathbb{C}^\wedge \setminus G)$.

<u>Proof</u>: Given $[\mu] \in M_0'(G)$, we would like to follow the example of the unit disk and consider

$$\lambda(w) = \int \frac{1}{w - z} d\mu(z).$$

This is certainly permissible for $w \notin G$, since then $\frac{1}{z - w} \in C(G)$. However we need to consider w in a neighborhood of $\mathbb{C}^\wedge \setminus G$. To do this we use the Tietze Extension Theorem to obtain a function $F_w \in C(G)$ such that $F_w(z) = \frac{1}{w - z}$ for all $z \in K$ where K is some compact set in G that supports μ. More precisely, let K be such a compact set and let $w \notin K$. Then $\frac{1}{z - w}\Big|_K$ is continuous on K and so may be extended to a continuous function F_w on G. We can then define

$$\lambda(w) = \int F_w(z)\,d\mu(z).$$

Because of our previous discussion, this integral is independent of the particular extension chosen and so we can simply write $\lambda(w) = \int \frac{1}{w - z}\,d\mu(z)$. What we have done, essentially, is prove the following.

(9.6) <u>Proposition</u>: <u>If</u> $\mu \in M_0(G)$ <u>and</u> μ <u>is supported by</u> K, <u>then there is an "extension"</u> μ' <u>of</u> μ <u>with</u> $\mu' \in C(K)^*$ <u>where</u> $C(K)$ <u>is the space of all continuous functions on</u> K <u>with the single norm</u> $\|\cdot\|_K$. <u>The connection is</u> $\int f\,d\mu = \int f|_K\,d\mu'$, $f \in C(G)$.

Thus, if $w_n \to w \notin K$ then $\lambda(w_n) \to \lambda(w)$ because $\frac{1}{w_n - z} \to \frac{1}{w - z}$ in $C(K)$. Similar reasoning makes it clear that λ is defined and holomorphic in $\mathbb{C}^\wedge\setminus K$ and $\lambda(\infty) = 0$, i.e. $\lambda \in H_0(\mathbb{C}^\wedge\setminus G)$.

So far we have a map $\Phi : M_0'(G) \to H_0(\mathbb{C}^\wedge\setminus G)$. What we want to do is show that it is one-one and onto, and in the process, compute its inverse. This requires a weak form of the Cauchy Integral Theorem, and a version of Fubini's theorem on the interchange of order in an "iterated integral".

In this first theorem, "oriented line segment" means a curve γ of the form $\gamma(t) = \gamma_0 + at$, $0 \le t \le 1$, where $a \in \mathbb{C}$. Informally it is a line segment with a specific "direction" imposed. If $\Gamma_1,\ldots,\Gamma_n$ are (oriented) line segments, then we write $\Gamma = \Gamma_1 \cup \Gamma_2 \cup\cdots\cup \Gamma_n$ and $\int_\Gamma f(z)dz$ to mean $\sum_{i=1}^{n} \int_{\Gamma_i} f(z)dz$.

(9.7) <u>Cauchy Theorem (Weak Form)</u>. <u>Given an open set</u> G <u>and a compact set</u> $K \subseteq G$, <u>there exists a finite sequence</u> $\Gamma_1,\Gamma_2,\ldots,\Gamma_n$ <u>of oriented line segments in</u> $G\setminus K$ <u>such that for each</u> $f \in H(G)$

a) $\displaystyle\int_\Gamma f(z)dz = 0$

b) $\displaystyle\frac{1}{2\pi i}\int_\Gamma \frac{f(z)}{z - a}\,dz = f(a)$ for each $a \in K$.

c) <u>If</u> $\lambda \in H_0(\mathbb{C}^\wedge\setminus G)$, <u>and</u> λ <u>is holomorphic in</u> $\mathbb{C}^\wedge\setminus K$, <u>let</u>

$$\lambda^*(w) = \frac{1}{2\pi i}\int_\Gamma \frac{\lambda(z)}{w-z}\,dz,\ w \in \mathbb{C}\setminus\Gamma.$$

<u>Then</u> $\lambda^* = \lambda$ <u>as elements of</u> $H_0(\mathbb{C}^\wedge\setminus G)$.

<u>Proof</u>: Let $\mathbb{Q} = \{Q_\nu\}$ denote a division of the plane $\mathbb{C}$ into solid squares Q_ν by equally spaced horizontal and vertical lines. Thus if $Q_\nu \in \mathbb{Q}$ then the corners of Q_ν are at $n\delta + im\delta$, $(n + 1)\delta + im\delta$, $n\delta + i(m + 1)\delta$ and $(n + 1)\delta + i(m + 1)\delta$ where n and m are integers and δ is the line spacing. We will think of Q_ν as the closed square. Choose the side length so $\delta < \frac{1}{2}\operatorname{dist}(K,\partial G)$. Let ∂Q_ν denote the boundary of Q_ν, which we view as a union of four oriented line segments with counterclockwise directions. Let $\mathbb{Q}^*$ be the collection of those $Q_\nu \in \mathbb{Q}$ that intersect K (i.e. some point of K lies inside Q_ν or on its boundary.) If two squares Q_α and Q_β in $\mathbb{Q}^*$ have an edge in common then it is necessarily oriented in opposite directions in ∂Q_α and ∂Q_β. We write $-L$ for an edge L with the reversed orientation. Let $\Gamma_1,\Gamma_2,\ldots,\Gamma_n$ denote all those (oriented) edges of squares in $\mathbb{Q}^*$ such that no $-\Gamma_i$ belongs to a square in $\mathbb{Q}^*$. (If we let $A = \bigcup_{Q_\nu\in\mathbb{Q}^*} Q_\nu$, then $\Gamma_1 \cup \Gamma_2 \cup\cdots\cup \Gamma_n$ is the topological boundary of A with orientations included.)

To prove (a) note that each Q_ν is entirely contained in G so

$$\int_{\partial Q_\nu} f(z)\,dz = 0.$$

Summing gives

$$\sum_{Q_\nu\in\mathbb{Q}^*}\int_{\partial Q_\nu} f(z)\,dz = 0. \tag{9.8}$$

But the left side of this equation is just

$$\sum_{i=1}^{n}\int_{\Gamma_i} f(z)\,dz = \int_\Gamma f(z)\,dz.$$

(Because all other edges of $\mathbb{Q}^*$ are common to two squares and so the integral in (9.8) over any such edge is canceled by an integral over the same edge in the reverse direction.) Thus (a) is proved.

To prove (b) observe that $a \in Q_\alpha$ for some square in $\mathbb{Q}^*$ and assume, for the moment, that a lies in the interior of Q_α. Then

$$f(a) = \frac{1}{2\pi i}\int_{\partial Q_\alpha} \frac{f(z)}{z-a}\,dz = \sum_{Q_\nu \in \mathbb{Q}^*} \frac{1}{2\pi i}\int_{\partial Q_\nu} \frac{f(z)}{z-a}\,dz$$

$$= \int_\Gamma \frac{f(z)}{z-a}\,dz.$$

The second equality follows because $f(z)/(z-a)$ is holomorphic on Q_ν, $\nu \neq \alpha$, and the third follows by the same argument as before.

Now, suppose a lies on an edge of Q_α. Then that edge cannot be one of the Γ_i and so both sides of

$$f(\zeta) = \frac{1}{2\pi i}\int_\Gamma \frac{f(z)}{z-\zeta}\,dz$$

are continuous in a neighborhood of a. Simply let $\zeta_n \to a$, $\zeta \in \operatorname{int} Q_\alpha$.

Finally we turn to part c. It suffices to show $\lambda^*(w) = \lambda(w)$ for w which do not lie in any of the squares of $\mathbb{Q}^*$. Let A_n be the block, centered at the origin, of $4n^2$ squares of $\mathbb{Q}$ that fit together to make a big square of side $2n\delta$, and let B_n be $A_n \setminus \mathbb{Q}^*$. Take n so large that both w and $\mathbb{Q}^*$ lie inside the boundary of A_n. If Q_0 is the square that contains w (by continuity we can suppose w is not on the edge of Q_0) we have

$$\lambda(w) = \frac{1}{2\pi i}\int_{\partial Q_0} \frac{\lambda(z)}{z-w}\,dz$$

and if Q_μ is any other square not in $\mathbb{Q}^*$ then

$$\frac{1}{2\pi i}\int_{\partial Q_\mu} \frac{\lambda(z)}{z-w}\,dz = 0.$$

Thus

$$\lambda(w) = \sum_{Q_\mu \in B_n} \frac{1}{2\pi i}\int_{Q_\mu} \frac{\lambda(z)}{z-w}\,dz.$$

If S_n is the large square of side $2n\delta$ that bounds A_n, we see from the above that

$$\text{(9.9)} \quad \lambda(w) = -\frac{1}{2\pi i}\int_\Gamma \frac{\lambda(z)}{z-w}\,dz + \frac{1}{2\pi i}\int_{S_n} \frac{\lambda(z)}{z-w}\,dz.$$

We claim that the last integral tends to 0 as $n \to \infty$. A simple estimate of its absolute value shows it is dominated by $\frac{4n\delta}{\pi(\delta n - |w|)} \sup\{|\lambda(z)| : z \in S_n\}$. This tends to 0 because $\lambda(z) \to 0$ as $z \to \infty$. Since the rest of (9.9) is independent of n, we see $\lambda(w) = \lambda^*(w)$ for w in a neighborhood of $\mathbb{C}\setminus G$. QED

The next result is a simple version of the Fubini theorem on interchanging the order of integration in an iterated integral.

(9.10) Fubini Theorem. Given two open sets G and G', measures $\mu \in M_0(G)$ and $\mu' \in M_0(G')$, and compact sets $K \subseteq G$ and $K' \subseteq G'$ which support, respectively, μ and μ', then

$$\int\left(\int F(z,z')d\mu(z)\right)d\mu'(z') = \int\left(\int F(z,z')d\mu'(z')\right)d\mu(z)$$

for all $F \in C(K \times K')$.

Proof: Notice that all the integrals make sense: For fixed z', $F(z,z') \in C(K)$ and $\int F(z,z')d\mu(z)$ is easily seen to be continuous on K' because of the uniform continuity of F on $K \times K'$. Because of the continuity of μ and μ' as functionals on $C(K)$ and $C(K')$ it suffices to prove the result for a set of functions dense in $C(K \times K')$. Now the polynomials in x, y, x' and y' ($z = x + iy$, $z' = x' + iy'$) are dense by the Stone-Weierstrass Theorem (2.29) and the equality is easy for such functions: it is enough to observe that

$$\int\left(\int f(z)g(z')\, d\mu(z)\right)d\mu'(z') =$$

$$\int f(z)d\mu(z) \cdot \int g(z')d\mu'(z') =$$

$$\int\left(\int f(z)g(z')d\mu'(z')\right)d\mu(z),$$

and that polynomials in x, y, x', y' are finite sums of such products. Thus, the result is proved. QED

(9.11) We return, at last, to the proof of the main duality theorem. We have, so far, a map

$$\Phi : H(G)^* \to H_0(\hat{\mathbb{C}}\setminus G)$$

which takes $L = [\mu]$ to $\lambda(w) = \int \frac{1}{w-z}\, d\mu(z)$. We want an inverse which will associate a measure to a $\lambda \in H_0(\hat{\mathbb{C}}\setminus G)$. So, given λ, take a compact set $K \subseteq G$ such that λ is holomorphic outside K, and let Γ be as in the weak Cauchy theorem. Then define $L_{\lambda,\Gamma}$ by

$$L_{\lambda,\Gamma}(f) = \frac{1}{2\pi i}\int_\Gamma f(w)\lambda(w)dw.$$

We must show that $L_{\lambda,\Gamma}$ is actually independent of Γ, provided Γ is chosen as in the weak Cauchy theorem relative to a compact set outside of which λ is analytic. To this end, we show

a) $L_{\Phi(L),\Gamma} = L$ for any $L \in H(G)^*$ provided K is chosen so that, in addition K supports μ for some extension μ of L to $C(G)$.

and

b) $\Phi(L_{\lambda,\Gamma}) = \lambda$ (as elements of $H_0(\mathbb{C}^\wedge \setminus G)$) for any $\lambda \in H_0(\mathbb{C}^\wedge \setminus G)$.

To get (a) observe, for $\lambda = \Phi(L)$, that

$$\begin{aligned} L_{\Phi(L),\Gamma}(f) &= \frac{1}{2\pi i}\int_\Gamma f(w)\lambda(w)dw \\ &= \frac{1}{2\pi i}\int_\Gamma f(w)\left[\int \frac{1}{w - z}\,d\mu(z)\right]dw \\ &= \int\left[\frac{1}{2\pi i}\int_\Gamma \frac{f(w)}{w - z}\,dw\right]d\mu(z) \\ &= \int f(z)d\mu(z) = L(f). \end{aligned}$$

Here we have used both Fubini's theorem and the weak Cauchy theorem. For (b) we have

$$\Phi(L_{\lambda,\Gamma})(w) = \int \frac{1}{w - z}\,d\mu(z)$$

where μ is an extension of $L_{\lambda,\Gamma}$. But the definition of $L_{\lambda,\Gamma}$ shows that

$$\mu = \frac{1}{2\pi i}\,\lambda(z)dz\big|_\Gamma$$

is a suitable extension. This means

$$\Phi(L_{\lambda,\Gamma})(w) = \frac{1}{2\pi i}\int \frac{1}{w - z}\,\lambda(z)dz = \lambda(w)$$

by the weak Cauchy theorem again.

Now we can prove $L_{\lambda,\Gamma}$ is independent of Γ. Let $\lambda \in H_0(\mathbb{C}^\wedge \setminus G)$ and let Γ and Γ' both satisfy the requirements to define $L_{\lambda,\Gamma}$ and $L_{\lambda,\Gamma'}$. This means there are compact sets K(resp. K') such that λ is holomorphic outside K(resp. K') and that Γ(resp. Γ') "surrounds"

K(resp. K') in the sense of the weak Cauchy theorem. Clearly then there is a single compact set K_S, containing $K \cup K' \cup \Gamma \cup \Gamma'$, outside of which λ is holomorphic. Choose a suitable Γ_S for this K_S. Then K_S satisfies the requirements for a). From b) it is clear that both $\Phi(L_{\lambda,\Gamma}) = \lambda = \Phi(L_{\lambda,\Gamma'})$. Applying (a) with Γ_S in place of Γ and $L_{\lambda,\Gamma}$ (resp. $L_{\lambda,\Gamma'}$) in place of L gives

$$L_{\lambda,\Gamma_S} = L_{\lambda,\Gamma}$$

(resp. $L_{\lambda,\Gamma_S} = L_{\lambda,\Gamma'}$).

We have thus proven the Main Duality Theorem, which we state here in its complete form.

(9.12) Main Duality Theorem. Define the map

$$\Phi : H(G)^* \to H_0(\mathbb{C}^\wedge \setminus G)$$

by $\Phi(L)(w) = L(\frac{1}{z - w})$. (The meaning of this is contained in the proof and the discussion preceding it.) The mapping Φ is linear, one-one and onto. Its inverse is given by

$$[\Phi^{-1}(\lambda)](f) = \frac{1}{2\pi i}\int_\Gamma f(w)\lambda(w)dw, \; f \in H(G),$$

where Γ is any curve in G satisfying the weak Cauchy theorem relative to a compact set $K \subseteq G$ such that λ is holomorphic in $\mathbb{C}^\wedge \setminus K$.

(9.13) Definition. The function $\Phi(L)$ is denoted $L^\wedge$ and is referred to as the Cauchy transform of L. $L^\wedge$ is defined as a germ in $H_0(\mathbb{C}^\wedge \setminus G)$ or as a function outside some compact set $K \subseteq G$. Analogously, if $\mu \in M_0(G)$ we write

$$\hat{\mu}(w) = \int \frac{1}{z - w}\, d\mu(z)$$

and call $\hat{\mu}$ the Cauchy transform of μ.

NOTES: See the references in the NOTES section of Chapter 6.

Exercises

1. Let $\{K_\alpha : \alpha \in A\}$ denote the set of all compacts in G that support $\mu \in M_0(G)$. Show that $K = \bigcap_{\alpha \in A} K_\alpha$ is non-empty and supports μ. (Show that the intersection of two supports is a support and then use compactness.)

2. Prove that if μ_1 and μ_2 are elements of $M_0(G)$ supported by disjoint sets then $\mu_1 \neq \mu_2$.

3. Is there a region $G \subseteq \mathbb{C}$ and a functional $L \in H(G)^*$ such that $L(z^n) = n!$, $n = 0,1,2,\ldots$? Such that $L(z^n) = n$?

4. Let A be a closed set in $\hat{\mathbb{C}}$. Discuss the dual of $H(A)$. A natural topology for $H(A)$ is obtained as follows. Write $A = \bigcap_{n=1}^{\infty} G_n$ where the G_n are open sets in $\hat{\mathbb{C}}$ with $G_n \supseteq \bar{G}_{n+1}$, $n = 1,2,\ldots$. Then $H(A)$ "=" $\bigcup_n H(G_n)$. Put on $H(A)$ the largest locally convex topology such that convergence in $H(G_n)$ for some n implies convergence in $H(A)$.

Another is to identify $H(A)$ with $H(G)^* \times \mathbb{C}$ for appropriate G and put the topology on $H(G)^*$ generated by sets of the form $U_f(L_0) = \{L \in H(G)^* : |L(f) - L_0(f)| < 1\}$ for $f \in H(G)$. (The weak*-topology. See Chapter 17.)

5. Let G_1 and G_2 be disjoint open sets in $\mathbb{C}$. Express $H(G_1 \cup G_2)^*$ in terms of $H(G_1)^*$ and $H(G_2)^*$. (We can identify $H(G_1 \cup G_2)$ with a subset of $H(G_i)$, $i = 1,2,$ in a natural manner.)

6. (See Exercise 5) $H(G_1)^*$ and $H(G_2)^*$ can be identified with subsets of $H(G_1 \cup G_2)^*$ in a natural manner. Show that if $H(G_1)^* = H(G_2)^*$ then $G_1 = G_2$.

7. Discuss $H(G)^*$ where $G = \{z : \text{Re } z \neq 0\}$. If f is analytic in the strip $\{z : |\text{Re } z| < \varepsilon\}$ and $\lim_{|z| \to \infty} f(z) = 0$, does f necessarily "belong to" $H(G)^*$?

8. a) Find a sequence $\lambda_n \to 0$ such that for no region G does there exist an $L \in H(G)^*$ such that $L(e^{nz}) = \lambda_n$, $n = 1,2,3,\ldots$.

b) Same question with $L(z^n) = \lambda_n$, $n = 1,2,3,\ldots$.

9. Prove Exercise 5, Chapter 8, from the Main Duality Theorem.

10. Produce if possible a region $G \subseteq \mathbb{C}$ and an $L \in H(G)^*$ such that

a) $L(z^n) = \frac{1}{n!}$

b) $L(z^n) = \frac{1}{n^2 + 1}$

c) $L(z^n) = n^3$

d) $L(z^n) = n^n$.

§10. Runge's Theorem

If $f \in H(G)$, G a connected open set, it is a consequence of the power series expansion for holomorphic functions that if $f(z_n) = 0$, $z_n \to z_0 \in G$, then $f = 0$ in G. It is also a consequence that if $f^{(n)}(z_0) = 0$ for $n = 0,1,2,\ldots$, then $f = 0$ in G. We adopt conventions about "sets with multiplicity" that allow us to treat both cases as one.

By a set with multiplicity in $\mathbb{C}^{\wedge}$ we mean a pair (E,m) consisting of a set $E \subseteq \mathbb{C}^{\wedge}$ and a "multiplicity function" $m : E \to \{1,2,3,\ldots\} \cup \{\infty\}$. By a limit point of (E,m) we mean an ordinary limit point of E or a point $e \in E$ with $m(e) = \infty$. For convenience we usually write E for (E,m) unless we wish to make explicit reference to m. Given a set with multiplicity (E,m), let $F = F(E)$ denote the following collection of functions: If $e \in E$, $e \neq \infty$ and $m(e) < \infty$, then $1/(z - e) \in F$; if $e \neq \infty$ and $m(e) = \infty$, then $1/(z - e)^k \in F$ for $k = 1,2,3,\ldots$; if $\infty \in E$ and $m(\infty) = \infty$, then $z^k \in F$ for $k = 0,1,2,\ldots$. Notice that nothing special occurs when $m(\infty) \neq \infty$.

(10.1) __Proposition.__ __Let__ $E \subseteq \mathbb{C}^{\wedge}\backslash G$, $\lambda \in H_0(\mathbb{C}^{\wedge}\backslash G)$, __and suppose__ (E,m) __has a limit point in each connected component of__ $\mathbb{C}^{\wedge}\backslash G$. __If, for each__ $e \in E$, λ __has a zero at__ e __(with infinite multiplicity when__ $m(e) = \infty$) __then it follows that__ $\lambda = 0$.

__Proof:__ Let U be an open set containing $\mathbb{C}^{\wedge}\backslash G$ in which λ is holomorphic. Without any loss of generality we may assume each connected component of U contains a component of $\mathbb{C}^{\wedge}\backslash G$. The result then follows from the two results in the first paragraph. QED

With the machinery at our disposal, Runge's Theorem becomes quite easy.

(10.2) __Runge's Theorem.__ __If__ $E \subseteq \mathbb{C}^{\wedge}\backslash G$ __is a set with multiplicity which has a limit point in each component of__ $\mathbb{C}^{\wedge}\backslash G$, __then the closed span of__ $F(E)$ __is__ $H(G)$.

__Proof:__ By the Hahn-Banach Theorem, it is enough to prove that if $L \in H(G)^*$ and $L(h) = 0$ for each $h \in F$, then $L = 0$. But if $\lambda(w) = L(\frac{1}{z - w})$ is the element of $H_0(\mathbb{C}^{\wedge}\backslash G)$ associated with L by the Main Duality Theorem, then λ has a zero at each $e \in E$ (of infinite multiplicity if $m(e) = \infty$). Thus $\lambda = 0$, whence $L = 0$. QED

This version of Runge's Theorem is stronger than the classical one. For example, if G is the unit disk and E = {1,2,3,...} we have proved that the rational functions with <u>simple</u> poles at the positive integers (and no pole at ∞) are dense in H(G). The proof we give yields this with no additional difficulty. The classical proof says only that the rational functions with poles at the positive integers (not necessarily simple poles) are dense in H(G). Moreover, the classical proof requires some modifications to obtain the stronger result. (They are, to be sure, not deep modifications.)

We remark here that Runge's Theorem provides an alternate proof that H(G) is separable: it says that linear span of F(E) is dense in H(G) whenever E is a set with multiplicity which has a limit point in each component of $\mathbb{C}^\wedge \setminus G$. It suffices to prove that E can be chosen to be countable. So let E consist of any countable dense subset of $\mathbb{C}^\wedge \setminus G$ and give each point of E infinite multiplicity. (Finite multiplicity would suffice except at isolated points of $\mathbb{C}^\wedge \setminus G$.)

(10.3) <u>Definition</u>. <u>A set</u> G ⊆ $\mathbb{C}^\wedge$ <u>is</u> <u>simply connected if</u> $\mathbb{C}^\wedge \setminus G$ <u>is</u> <u>connected</u>.

This is a simple working definition that is especially appropriate to analytic function theory. It is equivalent to the homotopy and homology definitions, but we will not be concerned with that here. A disadvantage of this definition is that it is not intrinsic (i.e. one must go outside of G) and so, unfortunately, it is not easy to prove that simple connectivity of plane sets is a topological property, i.e. invariant under homeomorphism. Our definition of simply connected is slightly non-standard in that it does not require G to be connected.

(10.4) <u>Corollary to Runge's Theorem</u>. <u>If</u> G ⊆ $\mathbb{C}$ <u>is simply connected</u> <u>then the polynomials are dense in</u> H(G).

<u>Proof</u>: Take E = {∞} with multiplicity m{∞} = ∞. Then the span of F(E) consists of the polynomials. QED

By using the same tools as in Theorem 8.4 (Hahn-Banach, Duality, Jensen's Theorem) and the above corollary the following result can be proved.

(10.5) <u>Theorem</u> <u>If</u> G <u>is a simply connected open subset of the open</u> <u>unit disk, and if</u> $\{\lambda_n\}$ <u>is a sequence of distinct complex numbers</u>

with

$$\limsup_{r\to\infty} \frac{1}{r}\int_0^r \frac{\lambda(t)}{t}\,dt \geq 1,$$

then the span of $\{e^{\lambda_n z}\}$ *is dense in* $H(G)$. (*Here* $\lambda(t) = \sum_{|\lambda_n|\leq t} 1$, *as before*.)

The proof is left as an exercise. It amounts to the same proof as Theorem 8.4, where the above corollary is invoked to show that a linear functional that annihilates z^n for all $n \geq 0$ must be 0.

We now illustrate a use of Runge's Theorem to construct examples. The following proposition is essentially trivial. Runge's Theorem will be used to show that an analogous result for more than one function is false.

(10.6) Proposition. *There is no* $f \in H(\mathbb{D})$ *such that* $\lim_{|z|\to 1} |f(z)| = +\infty$, *that is*

$$\lim_{r\to 1-} (\inf\{|f(z)| : r < |z| < 1\}) = \infty.$$

Proof: If there were such a function and it had no zeros, then on putting $g = 1/f$, we have $\lim_{|z|\to 1-} g(z) = 0$. This implies $g = 0$, which is impossible. On the other hand, if f has zeros it can have only finitely many and, on dividing by a polynomial with the same zeros and multiplicities, we return to the previous case. QED

(10.7) Proposition. *There exist two functions* $f, g \in H(\mathbb{D})$ *such that* $\lim_{|z|\to 1} |f(z)| + |g(z)| = +\infty$.

Proof: Let $0 < r_1 < r_2 < \cdots < 1$, $r_n \to 1$ be a sequence (to be chosen later) and let $f(z) = \sum_{j=0}^{\infty} n_j z^{n_j}$ (with the n_j also chosen later) such that $|f(z_n)| \to \infty$ for all sequences $\{z_n\}$ with $|z_n| = r_n$. Notice that also $\lim_{r\to 1-} f(r) = +\infty$. Since f is continuous on each circle $|z| = r_n$ and on the interval $[0,1)$, there is an open set $A \subseteq \mathbb{D}$ such that

i) A is the union of an open superset of $[0,1)$ and infinitely many disjoint annuli in $\mathbb{D}$, with one annulus containing each circle $\{|z| = r_n\}$

ii) $|f(z)| \to +\infty$ as $|z| \to 1$ in A.

iii) $\mathbb{D}\setminus A$ is a union of countably many disjoint simply connected annular sectors $\{S_i\}$.

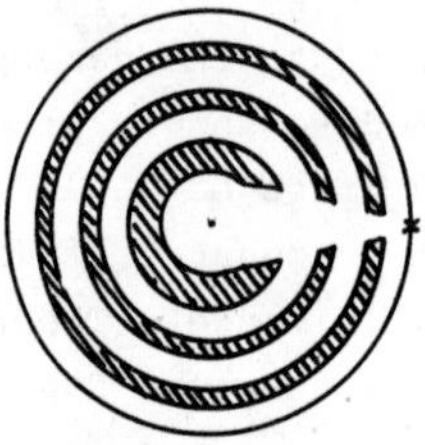

In the figure, the S_i are shaded, the remainder is A.

The proposition will be proved (assuming such a sequence r_n and function f can be constructed) if g can be constructed such that $|g(z)| \to \infty$ as $|z| \to 1$ through $\cup S_i$. This we do using Runge's Theorem. Let D^j be a disk which contains $S_1 \cup S_2 \cup \cdots \cup S_j$ but $(D^j)^-$ is disjoint from S_{j+1}. Define functions $g_1, g_2, \ldots$ by induction as follows. Let $g_1 \equiv 2$ and, having defined $g_1, g_2, \ldots, g_{n-1}$ define φ_n to be analytic in a neighborhood of $(D^n)^- \cup S_{n+1}$ so that it is 0 on D^n and equal to $n + 1 - \sum_{i=1}^{n-1} g_i$ on S_{n+1}. By Runge's Theorem there is a polynomial g_n such that $|g_n(z) - \varphi_n(z)| < 2^{-n}$ for $z \in D^n \cup S_{n+1}$. Let $g = \sum_{n=1}^{\infty} g_n$. (It is easily verified that this series converges in $H(\mathbb{D})$: on any disk D^j the series is eventually dominanted by a geometric series.) Now

$$g(z) = g_j(z) + \sum_{i=1}^{j-1} g_i(z) + \sum_{i=j+1}^{\infty} g_i(z).$$

The second sum is dominated on S_j by $\sum_{i=j+1}^{\infty} 2^{-i} = 2^{-j}$. Also $|g_j(z) + \sum_{i=1}^{j-1} g_i(z) - (j+1)| < 2^{-j}$, so $|g(z)| > j + 1 - 2^{-j} - 2^{-j} \geq j$ on S_j. Hence $\lim_{\substack{|z|\to 1 \\ z\in\cup S_j}} |g(z)| = +\infty$, as desired.

To obtain the sequence $\{r_n\}$ and the function f, start with $n_1 = 2$ and choose $r_1 < 1$ so that $n_1 r_1^{n_1} > 1$ and then choose m_1 so that

$$n_1 r_1^{n_1} > 1 + \sum_{j=m_1}^{\infty} j r_1^j$$

Having constructed $r_1 < r_2 < \cdots < r_k$; $n_1 < n_2 < \cdots < n_k$, and m_1, $m_2, \ldots, m_k$, choose $n_{k+1} > \max(m_k, n_k)$ such that

$$n_{k+1} > k + 1 + \sum_{j=1}^{k} n_j.$$

Then pick $r_{k+1} > r_k$ such that

$$n_{k+1} r_{k+1}^{n_{k+1}} > k + 1 + \sum_{j=1}^{k} n_j$$

and finally pick m_{k+1} so that

$$n_{k+1} r_{k+1}^{n_{k+1}} > k + 1 + \sum_{j=1}^{k} n_j + \sum_{j=m_{k+1}}^{\infty} j r_{k+1}^j.$$

Now, if $|z| = r_k$, then

$$\begin{aligned} |f(z)| &\geq n_k r_k^{n_k} - \sum_{j=1}^{k-1} n_j r_k^{n_j} - \sum_{j=k+1}^{\infty} n_j r_k^{n_j} \\ &\geq n_k r_k^{n_k} - \sum_{j=1}^{k-1} n_j - \sum_{j=m_k}^{\infty} j r_k^j \\ &> k. \end{aligned}$$

Thus $|f(z)| \to \infty$ if $|z| \to 1$ through the collection of circles $\bigcup_k \{z : |z| = r_k\}$. This is what we required. QED

NOTES: A constructive proof of Runge's Theorem can be found in many of the texts on complex analysis. It involves some of the same techniques already used in Chapter 9 (Theorem 9.7b) and the representation of a line integral as a limit of Riemann sums. See [Saks and Zygmund].

For some applications of Runge's Theorem see [Rubel, 1].

A theorem due to Szegö gives the degree of approximation possible in Runge's Theorem. See [Hille, vol. II, Theorem 16.6.5, pp. 304-305].

Exercises

1. Show that there exist $f, g \in H(\mathbb{D})$ such that for all m,n

$$\lim_{|z| \to 1} |f^{(m)}(z)| + |g^{(n)}(z)| = +\infty$$

(Hint: modify the construction of $\{r_k\}$ so that for each m,

$\lim_{k\to\infty} |f^{(m)}(z_k)| = +\infty$ when $|z_k| = r_k$ for all k.)

*2. Does there exist a continuous function φ on $\mathbb{D}$ such that $|f(z)| + |g(z)| \geq |\varphi(z)|$ is illegal when $f, g \in H(\mathbb{D})$?

*3. For any $n \geq 2$, do there exist n functions $f_1, f_2, \ldots, f_n \in H(\mathbb{D})$ such that for all $i,j = 1,2,\ldots,n$, $i \neq j$

(A) $|f_i(z)| + |f_j(z)| \to +\infty$ as $|z| \to 1$?

What about $|f_i(z)| + |f_j(z)| + |f_k(z)| \to +\infty$ if (A) doesn't work?

4. Do there exist two analytic functions $f, g \in H(\mathbb{D})$ such that $\lim_{|z|\to 1} |f(z)|\cdot|g(z)| = +\infty$?

5. An alternate version of Runge's Theorem is the following:

<u>Let</u> G <u>be an open set in</u> $\mathbb{C}$ <u>and</u> K <u>a compact subset of</u> G. <u>Let</u> $f \in H(G)$ <u>and</u> $\varepsilon > 0$. <u>Then there exists a rational function</u> r(z) <u>with poles off</u> G <u>such that</u> $\|f - r\|_K < \varepsilon$.

Give a proof of this along the following lines. Let μ be a measure supported by K such that $\int r\,d\mu = 0$ for all rational functions r with poles off G. Show that this implies $\int f\,d\mu = 0$ by writing $f(z) = \frac{1}{2\pi i}\int_\Gamma \frac{f(\zeta)}{\zeta - z}\,d\zeta$ and using Fubini's Theorem (9.10).

6. Given an entire function f(z), find a sequence $\{p_n\}$ of polynomials such that $\lim_{n\to\infty} p_n(0) = +\infty$ and $\lim_{n\to\infty} p_n(z) = f(z)$, $z \neq 0$.

7. Referring to the form of Runge's Theorem in Exercise 5, give a "constructive" proof along the following lines.

From $f(z) = \frac{1}{2\pi i}\int_\Gamma f(\zeta)(\zeta - z)^{-1}d\zeta$, where Γ is finite union of simple closed curves in G that "separates" K from $\hat{\mathbb{C}}\setminus G$ (see Chapter 9), approximate f by linear combinations of $(\zeta - z)^{-1}$ where $\zeta \in \Gamma$. Show that each $(\zeta - z)^{-1}$ can be approximated by a rational function with poles "closer" to $\hat{\mathbb{C}}\setminus G$ than ζ. That is, let ω be a path from ζ to $\hat{\mathbb{C}}\setminus K$ which does not meet K. Let ζ_0 satisfy $\zeta_0 \in \omega$, $|\zeta - \zeta_0| < \text{dist}(\zeta_0, K)$. Then

$$(\zeta - z)^{-1} = \sum_{n=0}^{\infty} \frac{(\zeta_0 - \zeta)^n}{(\zeta_0 - z)^{n+1}}$$

with uniform convergence on K. Similarly approximate $(\zeta_0 - z)^{-n-1}$ by rational functions with poles at ζ_1, where ζ_1 is further along ω. Continue.

§11. The Cauchy Theorem

Runge's Theorem can be used to prove Cauchy's Theorem. This will require the elements of integration theory described in Chapter 2. Recall that for a rectifiable curve $\gamma : [0,1] \to \mathbb{C}$, we let $\|\gamma\|$ denote its length, i.e. $\|\gamma\| = \int_0^1 |\gamma'(t)|dt$, so that

$$\left|\int_\gamma f(z)dz\right| \leq \|\gamma\| \cdot \|f\|$$

where $\|f\| = \sup\{|f(z)| : z \in \gamma^\wedge\}$. The reader is reminded that $\gamma^\wedge$ denotes the "physical curve", that is the image of γ.

(11.1) <u>Strong Cauchy Theorem</u>. <u>If G is a simply connected open set in $\mathbb{C}$, and γ is a closed rectifiable curve in G, then $\int_\gamma f(z)dz = 0$ for each $f \in H(G)$.</u>

<u>First Proof</u>. $\int_\gamma p(z)dz = 0$ for each polynomial p since each polynomial has a primitive in $\mathbb{C}$ (the primitive of z^n is $z^{n+1}/(n+1)$). Hence $\int_\gamma f(z)dz = 0$ for all $f \in H(G)$ since, by Runge's Theorem, f is the uniform limit on $\gamma^\wedge$ of a sequence of polynomials. QED

(11.2) <u>Definition</u>. <u>Given a set $K \subseteq G$ we define the hull of K with respect to G, K^G to be the following set: the union of K with all those components of $\mathbb{C}^\wedge \setminus K$ which lie entirely in G.</u>

Roughly speaking, K^G is formed from K by filling in the holes of K that lie in G. If G is simply connected, then K^G is the smallest simply connected superset of K (exercise). If K is compact, then K^G is compact since it is the complement in $\mathbb{C}^\wedge$ of the open set formed from all the components of $\mathbb{C}^\wedge \setminus K$ which meet $\mathbb{C}^\wedge \setminus G$. Two examples: if $K = \{z : |z| = \frac{1}{2}\}$ and $G = \{z : |z| < 1\}$, then $K^G = \{z : |z| \leq \frac{1}{2}\}$. If $G = \{z : 0 < |z| < 1\}$, then $K^G = K$.

<u>Second Proof</u>. (Using Fubini's Theorem) Let $K = (\gamma^\wedge)^G$ be the hull of $\gamma^\wedge$. Choose Γ by the <u>weak</u> Cauchy Theorem (9.7). For $z \in \gamma^\wedge$, $f(z) = \frac{1}{2\pi i}\int_\Gamma \frac{f(w)}{w-z}dw$. Thus, $\int_\gamma f(z)dz = \int_\gamma \frac{1}{2\pi i}\int_\Gamma \frac{f(w)}{w-z}dwdz = \int_\Gamma f(w)\frac{1}{2\pi i}\int_\gamma \frac{1}{w-z}dzdw$. The result will be proved if we can show $\frac{1}{2\pi i}\int_\gamma \frac{dz}{w-z} = 0$ for $w \notin K$. Letting $\varphi(w) = \frac{1}{2\pi i}\int_\gamma \frac{dz}{w-z}$, it is easy to see that $\varphi \in H_0(\mathbb{C}^\wedge \setminus K)$. But for w sufficiently large, say $|w| > 2\,\mathrm{dist}(0,\gamma^\wedge)$ we have

$$\frac{1}{w - z} = \frac{1}{w}\left(1 + \frac{z}{w} + \frac{z^2}{w^2} + \cdots\right)$$

with the series converging uniformly on $\gamma^\wedge$. Again, since $\int_\gamma z^n dz = 0$, we see $\varphi(w) = 0$ for such w. Since $\mathbb{C}^\wedge \setminus K$ is connected, $\varphi = 0$ in $\mathbb{C}^\wedge \setminus K$. QED

Before we continue we show that the system $\Gamma = \Gamma_1 \cup \Gamma_2 \cup \cdots \cup \Gamma_n$ obtained in the weak Cauchy theorem can be modified to obtain a curve Γ_0 which is again a union of oriented line segments but which can also be expressed as a disjoint union of finitely many simple closed curves. A few examples (which the reader may construct) convince one that the problem is to show that Γ_0 can be constructed so that at most two segments meet at each vertex. In the original construction of Γ the following configuration is possible:

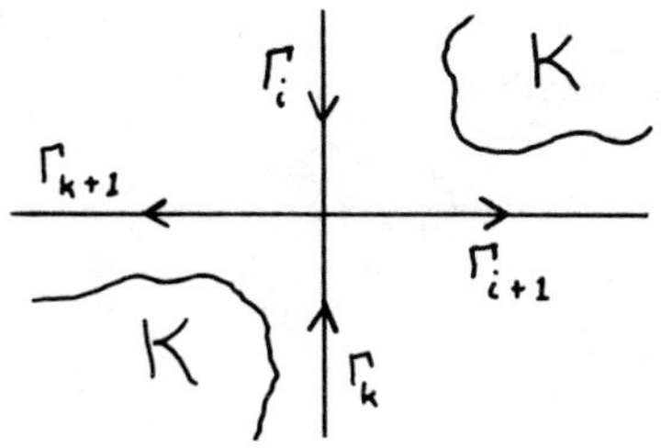

Figure 11.1

To avoid such a multiple point simply cut off the corners as pictured below.

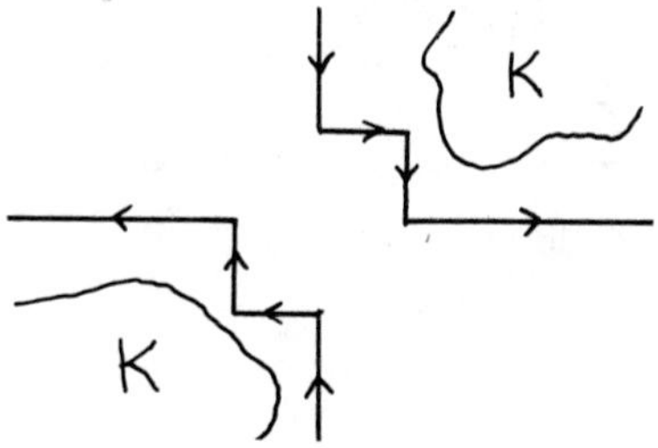

Figure 11.2

This is repeated for each such configuration and the resulting union of oriented segments is called Γ_0. To see that Γ_0 is a disjoint union of simple closed curves let $\Gamma_0 = \Gamma_1 \cup \Gamma_2 \cup \cdots \cup \Gamma_m$ with the Γ_i's chosen as follows. Let Γ_1 be any segment making up Γ_0, let Γ_2 be that (unique!) segment whose initial point is the terminal point of Γ_1, continue chaining Γ_2 to Γ_3, Γ_3 to Γ_4, etc., until Γ_k is reached whose terminal point is the initial point of Γ_1 (why is this possible?) Then $\Gamma_1 \cup \cdots \cup \Gamma_k$ is a simple closed curve (why?). Choose Γ_{k+1} to be any segment not already chosen and repeat the procedure. The result is, in a finite number of repetitions, a disjoint (why?) collection of simple closed curves.

(11.3) Definition. If γ is a closed rectifiable curve and $z \notin \gamma^\wedge$, then the winding number of γ around z, written $W(\gamma : z)$, is defined by

$$W(\gamma : z) = \frac{1}{2\pi i}\int_\gamma \frac{1}{w - z}\,dw.$$

Geometrically, $W(\gamma : z)$ is the number of times γ winds around z. Clearly, $W(\gamma : z)$ is in $H_0(\mathbb{C}^\wedge\setminus\gamma^\wedge)$. In each component of $\mathbb{C}^\wedge\setminus\gamma^\wedge$, $W(\gamma : z)$ is constant since $\frac{d}{dz}W(\gamma : z) = \frac{1}{2\pi i}\int_\gamma \frac{1}{(w - z)^2}\,dw = 0$. The integral is zero because $\frac{1}{(w - z)^2}$ has a primitive, namely $\frac{-1}{w - z}$, in a neighborhood of $\gamma^\wedge$. Actually, $W(\gamma : z)$ is an integer for each $z \notin \gamma^\wedge$. We will not prove this fact yet but the reader can easily use the weak Cauchy theorem to show it for the Γ obtained there. Clearly $W(\gamma : z) = 0$ in the component of $\mathbb{C}^\wedge\setminus\gamma^\wedge$ containing ∞.

(11.4) The Cauchy Integral Formula. If G is simply connected, and γ is a closed rectifiable curve in G, then for each $f \in H(G)$, and all $z \in G\setminus\gamma^\wedge$,

$$f(z)W(\gamma : z) = \frac{1}{2\pi i}\int_\gamma \frac{f(w)}{w - z}\,dw.$$

Proof: Consider

$$\frac{1}{2\pi i}\int_\gamma \frac{f(w)}{w - z}\,dw - f(z)W(\gamma : z) = \frac{1}{2\pi i}\int_\gamma \frac{f(w) - f(z)}{w - z}\,dw.$$

The function $F(w)$, which equals $\frac{f(w) - f(z)}{w - z}$ when $w \neq z$ and $F(z) = f'(z)$, is in $H(G)$. By the Cauchy Theorem (strong) $\int_\gamma F(w)\,dw = 0$, which gives the result. QED

Using Runge's Theorem, as in the first proof of Cauchy's Theorem, the following, stronger Cauchy Theorem can be proved.

(11.5) Very Strong Cauchy Theorem. A necessary and sufficient condition on the closed rectifiable curve γ, with $\gamma^\wedge \subseteq G$, such that $\int_\gamma f(z)\,dz = 0$ for each $f \in H(G)$, is that $W(\gamma : z) = 0$ for all $z \notin G$.

One direction is easy: if $z \notin G$ then $\frac{1}{w - z} \in H(G)$. We leave the details of the proof to the reader. The obvious analogue holds for finite unions of closed rectifiable curves, where, if $\gamma = \gamma_1 \cup \gamma_2 \cup \cdots \cup \gamma_n$, then $\int_\gamma$ means $\int_{\gamma_1} + \int_{\gamma_2} + \cdots + \int_{\gamma_n}$.

We turn to some applications of Cauchy's Theorem.

(11.6) Theorem. If G is simply connected and $f \in H(G)$, then f has a primitive $F \in H(G)$.

Proof: We can suppose that G is connected. Fix $z_0 \in G$ and define F by

$$F(z) = \int_{z_0}^{z} f(w)\,dw$$

where the integration is along any rectifiable curve which begins at z_0 and ends at z. The Cauchy Theorem says that F is well defined (the difference between the integrals defining F along two different curves is an integral around a closed curve) and if we differentate F "by hand" we get

$$\lim_{z' \to z} \frac{F(z') - F(z)}{z' - z} = \lim_{z' \to z} \frac{1}{z' - z} \int_{z'}^{z} f(w)\,dw$$

This limit is easily seen to be $f(z)$: take a straight line segment from z' to z as the path of integration and use the continuity of f at z. QED

An open set G for which all elements of $H(G)$ have primitives might be called "analytically simply connected". However, the following theorem shows that there is no difference between simply connected and analytically simply connected.

(11.7) Theorem: G is simply connected if and only if every $f \in H(G)$ has a primitive in $H(G)$.

One half of this has been proven. To prove the remaining half it suffices to exhibit a function in $H(G)$ which lacks a primitive when G is not simply connected. This requires the following result, which seems obvious intuitively.

(11.8) Lemma. If G is not simply connected then there exists a closed rectifiable curve γ and a point $w \notin G$, such that $\gamma^\wedge \subseteq G$ and $W(\gamma : w) = 1$.

Proof: $\mathbb{C}^\wedge \backslash G$ is not connected so it contains a clopen (in the relative topology it inherits from $\mathbb{C}^\wedge$) set K such that $K \neq \emptyset$ and $K \neq \mathbb{C}^\wedge \backslash G$. Then the set $H = \mathbb{C}^\wedge \backslash G \backslash K$ is also clopen. Since $\mathbb{C}^\wedge \backslash G$ is closed so also are H and K (in the topology of $\mathbb{C}^\wedge$). Since $H \cap K = \emptyset$ and $H \cup K = \mathbb{C}^\wedge \backslash G$ we can suppose $\infty \in H$ and K is a compact subset of $\mathbb{C}$. Let $G' = \mathbb{C}^\wedge \backslash H$; then G' is an open set in $\mathbb{C}$ which contains the compact set K. Construct Γ as in the weak Cauchy theorem, relative to G' and K. Then $\Gamma \subseteq G' \backslash K = G$ and $\frac{1}{2\pi i} \int_\Gamma \frac{1}{z - w}\, dz = 1$ for $w \in K$. As we saw earlier, Γ can be taken to be a disjoint union of simple closed curves. If we examine the proof of the weak Cauchy theorem, we see that if γ is one of these simple closed curves, then $\frac{1}{2\pi i} \int_\gamma \frac{1}{z - w}\, dz$ is either 0 or 1 when $w \in K$. Fix $w \in K$ and pick γ so that $\frac{1}{2\pi i} \int_\gamma \frac{1}{z - w}\, dz = 1$. This is the result we needed. QED

Remark: It requires a little more of a topological argument, but w can be preassigned in any bounded component of $\mathbb{C}^\wedge \backslash G$ for this lemma. The topology required is the following: If X is a compact Hausdorff space and a, b are two points in different components of X, then there is a clopen set containing a but not b.

(11.9) Proof of Theorem. If G is not simply connected let w and γ be as in the lemma. Then $\frac{1}{z - w} \in H(G)$ but $\frac{1}{z - w}$ cannot have a primitive because $\int_\gamma \frac{1}{z - w}\, dz \neq 0$. QED

(11.10) The Algebra $A(K)$. Here, K is a compact set in $\mathbb{C}$, and $A(K)$ consists of all functions that are continuous on K and holomorphic in int K. For $f \in A(K)$, we define $\|f\| = \sup\{|f(z)| : z \in K\}$. It is clear that $A(K)$ is a subalgebra of $C(K)$, and, since uniform limits of holomorphic functions are holomorphic, that $A(K)$ is closed in $C(K)$.

(11.11) Theorem: Let G be an open set in $\mathbb{C}$ and K a compact subset of G. Let Γ be a finite union of closed rectifiable curves in G such that $W(\Gamma : z) = 0$ when $z \notin G$ and $W(\Gamma : z) = 1$ when $z \in K$. Let f be a function in $H(G)$ all of whose zeros lie in K. Then $\frac{1}{2\pi i} \int_\Gamma \frac{f'(z)}{f(z)}\, dz$ equals the number of zeros of f in K, counted according to multiplicity.

Proof: Let $z_1, \ldots, z_m$ be the zeros of f with multiplicities $k_1, \ldots, k_m$. Then $\frac{f'(z)}{f(z)} - \sum \frac{k_j}{z - z_j} \in H(G)$, so by the Very Strong

Cauchy Theorem (and the remark following it) its integral on Γ is zero. Thus

$$\frac{1}{2\pi i}\int_\Gamma \frac{f'(z)}{f(z)}\,dz = \sum \frac{1}{2\pi i}\int_\Gamma \frac{k_j}{z - z_j}\,dz$$

$$= \sum k_j\, W(\Gamma : z_j)$$

$$= \sum k_j,$$

and this sum is what is meant by "according to multiplicities." QED

(11.12) Rouché's Theorem. *If Φ and Ψ are both functions in* $A(J)$ *for a compact set* J, *and if* $|\Phi(z)| < |\Psi(z)|$ *for* $z \in \partial J$, *then* Ψ *and* $\Psi + \Phi$ *have the same number of zeros in* J.

Proof: Notice that Ψ can have no zeros on ∂J and the same is true of $\Psi + \Phi$. Let $K = \{z \in J : |\Phi(z)| \geq |\Psi(z)\}$ so that K is a compact set in int J. For each t with $0 \leq t \leq 1$, let $F_t = \Psi + t\Phi$. Now, the zeros of F_t lie in K since $F_t(z) = 0$ implies $|\Psi(z)| = t|\Phi(z)|$. Choosing Γ in $(\text{int } J)\setminus K$ by the weak Cauchy Theorem, the number of zeros of F_t, $n(F_t)$, is

$$n(F_t) = \frac{1}{2\pi i}\int_\Gamma \frac{F_t'(z)}{F_t(z)}\,dz.$$

This formula makes it clear that $n(F_t)$ is an integer valued continuous function of t on $[0,1]$. As $[0,1]$ is connected, we conclude $n(F_t)$ is constant; in particular $n(F_0) = n(F_1)$. Since $F_0 = \Psi$ and $F_1 = \Psi + \Phi$, this is the required result.

As corollaries we obtain three versions of Hurwitz's Theorem.

(11.13) Hurwitz's Theorem. (1) *Let* K *be compact, and let* f, $g \in A(K)$. *If* $f(z) \neq 0$ *for* $z \in \partial K$ *then* f *and* g *have the same number of zeros in* K *provided that* $\|f - g\| < \inf\{|f(z)| : z \in \partial K\}$. *(We count multiplicities.)*

(2) *Let* f *be as in* (1) *and let* $\{f_k\}$ *be a sequence in* $A(K)$ *converging to* f *in the sense of the norm* $\|\cdot\|$. *Then there is a* k_0 *such that* $n(f_k) = n(f)$ *for* $k \geq k_0$.

(3) *If* $f_n \to f$ *in* $H(G)$ *and if* f *is not identically zero in any component of* G *and* f *has at least* N *zeros in* G, *then there is an* n_0 *such that for* $n \geq n_0$, f_n *has at least* N *zeros in* G.

Proof: All follow from form (1), which follows directly from Rouché's Theorem with $\Psi = f$, $\Phi = g - f$. QED

(11.14) Definition. An analytic function is called univalent if it is one-one. The term schlicht is sometimes used with the same meaning.

A direct consequence of Hurwitz (3) is the following.

(11.15) Theorem. If G is connected and $\{f_n\}$ is a sequence of univalent functions in $H(G)$ that converges in $H(G)$ to a function f, then f is either univalent or constant.

Proof: Apply the third version of Hurwitz' Theorem to $f - f(z_0)$ to see that $f - f(z_0)$ must have at most one zero, if it is not a constant function. QED

Rouché's Theorem has a stronger form (see Exercise 5) which form has a sort of a converse. The key observation that leads to the stronger form is that we only need enough hypotheses to insure that F_t has no zeros on ∂J. If we let $F_t = tf + (1 - t)g$ then f and g have the same number of zeros in $\operatorname{int} J$ provided f and g are not zero on ∂J and that, for each $z \in \partial J$ the line segment from $f(z)$ to $g(z)$ does not meet 0. This will be the case if

$$|f(z) - g(z)| < |f(z)| + |g(z)|, \quad z \in \partial J.$$

Now we specialize to $J = \mathbb{D}^-$, the closed unit disk. If $a \in \mathbb{D}$ then the rational function $B(z) = (z - a)/(1 - \bar{a}z)$ has one zero in $\mathbb{D}$ and $|B(e^{i\theta})| = |e^{i\theta}(e^{i\theta} - a)/e^{-i\theta} - \bar{a})| = 1$. Thus $|B(z)f(z)| = |f(z)|$ for $z \in \partial\mathbb{D}$. Now we can generalize Rouché's Theorem to the following.

(11.16) Let $f, g \in A(\mathbb{D}^-)$ and let $\alpha(z)$ and $\beta(z)$ have the form

$$(*) \qquad \begin{cases} \alpha(z) = cz^m \displaystyle\prod_{i=1}^{n} \frac{z - a_i}{1 - \bar{a}_i z} \\ \\ \beta(z) = dz^k \displaystyle\prod_{j=1}^{\ell} \frac{z - b_i}{1 - \bar{b}_j z} \end{cases}$$

where $|c| = |d| = 1$, $|a_i|$, $|b_j| < 1$. Suppose that α and β

have the same number of zeros in $\mathbb{D}$ *and that for all* $z \in \partial\mathbb{D}$ $|\alpha(z)f(z) + \beta(z)g(z)| < |f(z)| + |g(z)|$. *Then* f *and* g *have the same number of zeros in* $\mathbb{D}$.

In this form Rouché's Theorem has a converse.

(11.17) *Theorem.* (Challener, Rubel) *Suppose* f *and* g *are in* $A(\mathbb{D}^-)$, *have no zeros on* $\partial\mathbb{D}$ *and have the same number of zeros in* $\partial\mathbb{D}$. *Then there exist* $\alpha(z)$ and $\beta(z)$ *of the form* (*) *such that* α *and* β *have the same number of zeros and*

$$|\alpha(z)f(z) + \beta(z)g(z)| < |f(z)| + |g(z)|.$$

for $z \in \partial\mathbb{D}$.

(11.18) *Lemma.* *Let* γ *be continuous on* $\partial\mathbb{D}$ *and* $|\gamma(z)| = 1$ *for all* $z \in \partial\mathbb{D}$. *If* $\varepsilon > 0$, *then there exist functions* α, β *of the form* (*) *such that* $|\alpha(z) - \beta(z)\gamma(z)| < \varepsilon$ *for* $z \in \partial\mathbb{D}$.

Proof: (Sketch) By the Stone-Weierstrass Theorem (2.29), $\gamma(e^{i\theta})$ can be uniformly approximated by sums of the form $\sum_{-n}^{n} c_k e^{ik\theta} = \sum_{-n}^{n} c_k z^k$ where $z = e^{i\theta}$. Thus there is a rational function $r(z)$ such that $|\gamma(z) - r(z)| < \varepsilon$ for $z \in \partial\mathbb{D}$. Let the zeros of r be a_i, $i = 1,2,3,\ldots,k$ and $1/\bar{c}_j$, $j = 1,2,\ldots,\ell$ where $|a_i|, |c_j| < 1$. Similarly denote the poles by b_i, $i = 1,2,\ldots,m$ and $1/\bar{d}_j$, $j = 1,2,\ldots,n$, $|a_i|, |d_j| < 1$. Then

$$r(z) = c\,\frac{\prod_{i=1}^{k}(z - a_i)\prod_{j=1}^{\ell}(1 - \bar{c}_j z)}{\prod_{i=1}^{m}(z - b_i)\prod_{j=1}^{n}(1 - \bar{d}_j z)}.$$

Using the power series for $\log(1 - \bar{a}_i z)$ we can write $1 - \bar{a}_i z = \exp h_i$ for some h_i analytic in a neighborhood of $\mathbb{D}^-$. The same goes for $1 - \bar{b}_i z$, $1 - \bar{c}_i z$ and $1 - \bar{d}_i z$. This gives

$$r(z) = z^p \frac{\alpha_0(z)}{\beta_0(z)} e^{h(z)}$$

where p is the order of the pole or zero of r at the origin and α_0 and β_0 have the form (*). Since $r(e^{i\theta})$ is very nearly 1,

$\mathrm{Re}\,h(e^{i\theta})$ is nearly zero. If we can approximate $\exp i \,\mathrm{Im}\, h$ by the appropriate quotient, we will be done. Let $v = \exp i\,\mathrm{Im}\,\frac{h}{2}$. Again v can be approximated by a rational function $s(z)$. Now $s(z)/\overline{s(1/\bar{z})}$ will be of the form (*) and, for $|z| = 1$,

$$s(z)/\overline{s(1/\bar{z})} = s(z)/\overline{s(z)} = s(z)^2/|s(z)|^2$$

Now s^2 approximates $\exp i\,\mathrm{Im}\,h$ and $|s|^2$ approximates 1, so $s(z)/\overline{s(1/\bar{z})}$ approximates $\exp i\,\mathrm{Im}\,h$. QED

(11.18) Lemma. <u>Given $f, g \in A(\mathbb{D}^-)$ having no zeros on $\partial\mathbb{D}$, there exist α and β of the form (*) such that</u>

$$|\alpha f + \beta g| < \max\{|f|,|g|\} \text{ on } \partial\mathbb{D}.$$

Proof: Consider $h = \frac{g}{f}$ and $\gamma(e^{i\theta}) = \dfrac{h(e^{i\theta})}{|h(e^{i\theta})|}$. Then we may apply Lemma 11.17 to $-\gamma$ to obtain

$$\left|-\gamma - \frac{\alpha}{\beta}\right| < \frac{\min\limits_{|z|=1} \{|g(z)|,|f(z)|\}}{\max\limits_{|z|=1} |f(z)|}$$

on $|z| = 1$. Thus, on $|z| = 1$,

$$\begin{aligned}
|\beta g + \alpha f| &= \left|\frac{g}{f} + \frac{\alpha}{\beta}\right| |f| = \left|h + \frac{\alpha}{\beta}\right| |f| \\
&= \left|\gamma|h| + \frac{\alpha}{\beta}\right| |f| \\
&\le |f| \left(\left|\gamma + \frac{\alpha}{\beta}\right| + |1 - |h||\right) \\
&\le |f| \left(\frac{\min\limits_{|z|=1} \{|g(z)|,|f(z)|\}}{\max\limits_{|z|=1} |f(z)|} + |1 - |h||\right) \\
&\le \min_{|z|=1} \{|g(z)|,|f(z)|\} + ||f| - |g|| \\
&\le \max\{|g|,|f|\}. \quad \text{QED}
\end{aligned}$$

Proof of Theorem. From the above Lemma there exist α and β such that

$$|\alpha f + \beta g| < \max\{|f|,|g|\} \le |\alpha f| + |\beta g|.$$

From (11.16) we see that αf and βg have the same number of zeros in $\mathbb{D}$. Since f and g are assumed to have the same number of zeros, it follows that α and β have the same number as well. QED

NOTES: To see that $W(\gamma : z)$ is always an integer, fix γ and z_0. Cover $\gamma^\wedge$ by disks not containing z_0. In each disk $(w - z_0)^{-1}$ has a primitive, namely a branch of $\log(w - z_0)$. Thus there are points $w_0, w_1, \ldots, w_n = w_0$ such that

$$W(\gamma : z_0) = \sum_{i=0}^{n-1} \frac{1}{2\pi i} \int_{w_i}^{w_{i+1}} \frac{1}{w - z_0}\, dw$$

$$= \frac{1}{2\pi i} \sum_{i=1}^{n} F_i(w_{i+1}) - F_i(w_i)$$

where each F_i is a branch of $\log(w - z_0)$. Since different branches differ by integer multiples of $2\pi i$, $W(\gamma, z_0)$. is an integer.

Version 11.16 of Rouché's Theorem can be found in [T. Esterman] and in [I. Glicksberg]. Theorem 11.17 is the content of the paper [D. Challener and L. A. Rubel].

Exercises

1. Prove that the procedure outlined in 11.2 yields a finite union of simple closed curves.

2. Let Γ be the curve obtained in the weak Cauchy Theorem. Prove that $W(\Gamma : z)$ is an integer for each $z \notin \Gamma^\wedge$. Prove in fact that $W(\Gamma : z)$ is zero or one.

3. Prove the statement at the end of the proof of 11.6:

$$\lim_{z' \to z} \frac{1}{z' - z} \int_{z'}^{z} f(w)\, dw = f(z)$$

whenever f is continuous and the integral is taken over a straight line segment from z' to z.

4. Verify the Remark following Lemma 11.8.

5. Use Rouché's Theorem to prove that every polynomial of degree n

has n zeros. (Hint: Compare $|\sum_{n=0}^{k-1} a_n z^n|$ with $|a_k||z^k|$ on a large circle about the origin.)

6. Observe that the proof of (11.12) Rouché's Theorem hinges on the fact that $\Psi + t\Phi$ has no zeros in a neighborhood U of ∂J for any t, $0 \leq t \leq 1$. This can be seen geometrically by observing that for each $z \in \partial J$ and $t \in [0,1]$, the complex number $\Psi(z) + t\Phi(z)$ lies on the line segment from $\Psi(z)$ to $\Psi(z) + \Phi(z)$. The latter

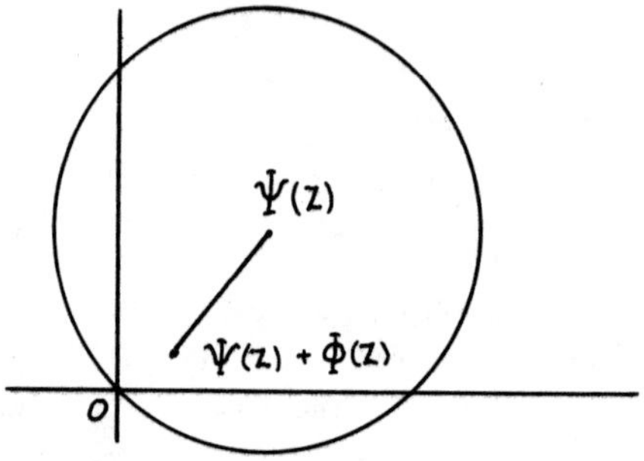

Figure 11.3

point lies in the disk about $\Psi(z)$ of radius $|\Psi(z)|$, a disk that does not contain 0.

Use a similar geometric picture to show the following generalization of Rouché's Theorem ([Estermann] and [Glicksberg]).

<u>If</u> $f, g \in A(J)$ <u>and</u> $|f - g| < |f| + |g|$ <u>on</u> ∂J, <u>then</u> f <u>and</u> g <u>have the same number of zeros in</u> J.

Hint: The condition $|f(z) - g(z)| < |f(z)| + |g(z)|$ implies that the line segment from $f(z)$ to $g(z)$ does not contain 0, i.e. $F_t(z) \equiv tf(z) + (1 - t)g(z)$ is never zero for any $z \in \partial J$ and $0 \leq t \leq 1$.

7. Show that if $r(z)$ is a rational function such that $|r(z)| = 1$ for $z \in \partial\mathbb{D}$, then $r(z) = \alpha(z)/\beta(z)$ for α and β satisfying (*) of (11.16).

8. a) Show that all five roots of the equation $z^5 + 15z + 1 = 0$ lie in $|z| < 2$ but that only one lies in $|z| < \frac{3}{2}$.

b) Show that $ze^{a-z} = 1$ has precisely one root in $|z| \leq 1$ when $a > 1$. Explain why this root must be positive. (Compare e^{-a} and ze^{-z} on $|z| = 1$).

c) Use Rouché's Theorem to show that $f(z) = \prod_{i=1}^{m} \frac{z - a_i}{1 - \bar{a}_i z}$ has the same number of zeros as $f - a$, when $|a| < 1$ and $|a_i| < 1$, $i = 1,2,\ldots,m$.

d) Given $r > 0$ show that there is an integer $m(r)$ such that all roots of

$$1 + z + \frac{z^2}{2!} + \cdots + \frac{z^m}{m!} = 0$$

belong to $|z| > r$ for all $m > m(r)$.

e) Let $f \in H(\mathbb{D})$ and let $0 < r < 1$. Let $n(r)$ be the number of zeros of f in $|z| < r$. Let $f(z) = \sum_{k=0}^{\infty} a_k z^k$. Show that

$$\inf_{|z|<r} |f(z)| \leq |a_0| + |a_1| + \cdots + |a_{n(r)}|.$$

How can this be improved (that is, the right hand side made smaller)?

9. Construct a closed rectifiable curve γ so that $W(\gamma : \frac{1}{n}) \to +\infty$ as $n \to \infty$.

10. Prove the maximum modulus theorem for $f \in H(\mathbb{D})$ by the idea

$$[f(z)]^n = \frac{1}{2\pi i} \int \frac{[f(z)]^n}{\zeta - z} d\zeta$$

$$|f(z)| \leq K^{1/n}\left\{\int |f(\zeta)|^n d|\zeta|\right\}^{1/n}$$

etc.

*11. If $f(z) = |z|$, is there any simple closed curve γ and point z_0 surrounded by γ such that

$$f(z_0) = \frac{1}{2\pi i} \int_\gamma \frac{f(z)}{z - z_0} dz?$$

§12. Constructive Function Theory

The goal of this section is the construction, by means of sums and products of simpler functions, of holomorphic functions with prescribed behavior. In what follows, when we speak of a sequence of complex numbers we ordinarily mean a sequence with multiplicity, so that a function taking some value at a point of the sequence must take that value with the appropriate multiplicity. We also exclude the trivial function $f \equiv 0$ unless otherwise noted.

(12.1) Definitions. (Region) A region is a connected open set. (Admissible) A sequence W in a region G is admissible if it has no limit point in G.

We will use $\mathcal{Z}(f)$ to denote the sequence of zeros of a function $f \in \mathcal{H}(G)$, with multiplicity.

(12.2) Theorem (Weierstrass). Given a region G and a sequence W in G, then there is an $f \in \mathcal{H}(G)$ with $\mathcal{Z}(f) = W$ if and only if W is admissible.

Proof: The "only if" direction has been done in Chapter 2. (The Uniqueness Theorem (2.15).) We require some preliminary material on infinite products. The basic idea of the proof is to find functions f_n such that f_n vanishes at one point of W and nowhere else and multiply these functions together. The crucial problem is to make sure that this (infinite) product doesn't become zero anywhere except on W.

(12.3) Definition. The infinite product $\prod_{k=1}^{\infty} u_k$ converges to u, written $\prod u_k = u$, provided (i) at most finitely many of the u_n are zero (say $u_k \neq 0$ for $k \geq n_0$), (ii) the sequence $\prod_{k=n_0}^{n} u_k$ converges to some number $u' \neq 0$, and (iii) $u = \left(\prod_{k=0}^{n_0-1} u_k\right) \cdot u'$.

(12.4) Definition. Let $u_k : E \to \mathbb{C}$ be functions on some set E. We say that $\prod u_k = u$ uniformly on E provided (i) There exists some n_0 such that $u_n(x) \neq 0$ for all $n \geq n_0$ and all $x \in E$, (ii) The sequence $\prod_{k=n_0}^{n} u_k(x)$ converges uniformly to some

$u'(x)$, <u>with</u> $u'(x) \neq 0$ <u>for all</u> $x \in E$ <u>and</u> (iii)

$$u(x) = \left(\prod_{k=1}^{n_0-1} u_k(x)\right) u'(x).$$

A word of explanation is in order about these definitions. The reason for excluding zero so many places is to ensure that the limit u is zero only when some factor in the product is zero. For infinite products 0 must be avoided almost as ∞ is avoided in infinite sums. In fact, one frequently hears "divergence to 0" to indicate that an infinite product has partial products which are all nonzero, but which tend to 0.

Convergence of an infinite product in $C(G)$ is defined to mean uniform convergence on compact subsets of G. It is easy to prove that if each u_k is in $H(G)$ and $\prod u_k = u$ (convergence in $C(G)$) then $u \in H(G)$. Moreover, if no u_k is identically zero then u is not identically zero.

(12.5) <u>Cauchy Criterion</u>. (Ratio form) $\prod u_k$ <u>is convergent if and only if</u> $\lim_{m,n\to\infty} \prod_{k=m}^{n} u_k = 1$, $\prod u_k(x)$ <u>converges uniformly if and only if</u> $\lim_{m,n\to\infty} \prod_{k=m}^{n} u_k(x) = 1$ <u>uniformly</u>

<u>Proof</u>: We prove only the first part. Suppose $\lim_{m,n\to\infty} \prod_{k=m}^{n} u_k = 1$. Then condition (i) of (12.3) is clear: choose n_0 so large that $\left|\prod_{k=n_0}^{n} u_k - 1\right| < \frac{1}{2}$ for all $n \geq n_0$. With this n_0 we have, for $n > m$,

$$\left|\prod_{k=n_0}^{n} u_k - \prod_{k=n_0}^{m-1} u_k\right| = \left|\prod_{k=n_0}^{m-1} u_k\right| \left|\prod_{k=m}^{n} u_k - 1\right| \to 0$$

as $n,m \to \infty$. Thus $\prod_{k=n_0}^{n} u_k$ is a Cauchy sequence and converges to some u' with $|u' - 1| < \frac{1}{2}$. Thus (ii) is satisfied and so is (iii). Conversely, suppose (12.3) is satisfied and let $\varepsilon > 0$. Choose n_1 so large that $\left|\prod_{k=n_0}^{n} u_k - u'\right| < \varepsilon|u'|$ for all $n \geq n_1$. Then if $n > m > n_1$,

$$\left|\prod_{k=m}^{n} u_k - 1\right| \le \left(\prod_{k=n_0}^{m-1} u_k\right)^{-1} \left|\prod_{k=n_0}^{n} u_k - \prod_{k=n_0}^{m-1} u_k\right|$$

$$\le 2\varepsilon |u'| / \left(\prod_{k=n_0}^{m-1} u_k\right)$$

$$\le \frac{2\varepsilon}{1-\varepsilon}$$

since $\left|\prod_{k=n_0}^{m-1} u_k\right| > (1-\varepsilon)|u'|$. QED

In the following let $a_k = u_k - 1$, so that $u_k = 1 + a_k$.

(12.6) Definition. $\prod u_k$ is absolutely convergent provided $\prod(1 + |a_k|)$ converges.

(12.7) Proposition. If $\prod u_k$ is absolutely convergent, then it is convergent.

Proof: First observe that $\prod_m^n (1 + a_k) - 1$ can be expanded as a sum $\sum \prod a_k$ of all possible products of distinct elements of $\{a_m, a_{m+1}, \ldots, a_n\}$. Thus

$$\left|\prod_m^n u_k - 1\right| = \left|\prod_m^n (1 + a_k) - 1\right|$$

$$= \left|\sum \prod a_k\right| \le \sum \prod |a_k|$$

$$= \prod_m^n (1 + |a_k|) - 1,$$

so $\prod u_k$ converges by the Cauchy criterion. QED

(12.8) Proposition. Any rearrangement of an absolutely convergent product is absolutely convergent to the same value.

Proof: This depends on the fact that $\prod (1 + |a_k|)$ converges if and only if its partial products are bounded. Details are left as an exercise. QED

Remark: If $\prod(1 + |a_k|)$ is uniformly convergent (or convergent in $C(G)$) then $\prod (1 + a_k)$ is uniformly convergent (respectively, convergent in $C(G)$). This is obtained from the proof that absolute convergence implies convergence by adding the word "uniform" in front of "Cauchy criterion".

(12.9) Definition. If λ is a non-negative integer, we define $E(z,\lambda)$, the Weierstrass primary factor of order λ by

$$E(z,0) = 1 - z$$

and

$$E(z,\lambda) = (1 - z)\exp(z + \frac{z^2}{2} + \cdots + \frac{z^\lambda}{\lambda}) \quad \text{if} \quad \lambda > 0.$$

Notice that $z + \frac{z^2}{2} + \cdots + \frac{z^\lambda}{\lambda}$ is the λth partial sum of $\log \frac{1}{1 - z}$ so that $E(z,\lambda)$ converges to 1 in $H(\mathbb{D})$ as $\lambda \to \infty$, where $\mathbb{D} = \{z : |z| < 1\}$.

(12.10) Proposition. If $|z| \leq \frac{1}{2}$, then

$$|E(z,\lambda) - 1| \leq 4|z|^{\lambda+1}.$$

Proof: For $|z| < 1$ we have

$$|E(z,\lambda) - 1| = |(1 - z)\exp(z + \frac{z^2}{2} + \cdots + \frac{z^\lambda}{\lambda}) - 1|$$

$$= |\exp(-\frac{z^{\lambda+1}}{\lambda + 1} - \frac{z^{\lambda+2}}{\lambda + 2} - \cdots) - 1|.$$

Since $|e^u - 1| \leq e^{|u|} - 1$ (look at the power series!) we have (writing $r = |z|$)

$$|E(z,\lambda) - 1| \leq \exp(\frac{r^{\lambda+1}}{\lambda + 1} + \frac{r^{\lambda+2}}{\lambda + 2} + \cdots) - 1$$

$$\leq \exp(r^{\lambda+1} + r^{\lambda+2} + \cdots) - 1$$

$$= \exp(\frac{r^{\lambda+1}}{1 - r}) - 1$$

$$\leq \exp 2r^{\lambda+1} - 1$$

The last inequality comes from $r \leq \frac{1}{2}$. Now, because $\exp x - 1 \leq 2x$ when $0 \leq x \leq \frac{1}{2}$ and $2r^{\lambda+1} \leq \frac{1}{2}$ if $\lambda > 0$, we see that $\exp 2r^{\lambda+1} - 1 < 4r^{\lambda+1}$. In case $\lambda = 0$, the result is trivial. QED

The factor 4 in this result can easily be improved, but this is not important.

(12.11) Proof of Theorem (12.2) in the special case $G = \mathbb{C}$. Here $W = \{w_n\}$ is admissible if and only if $\lim_{n\to\infty} w_n = \infty$. Define f by

$$f(z) = z^k \prod_{w_n \neq 0} E(\frac{z}{w_n}, \lambda_n)$$

where k is the multiplicity of 0 if $0 \in W$, $k = 0$ otherwise, and the λ_n are chosen so that the product converges in $H(\mathbb{C})$. This can be done as follows. We need only show uniform convergence on each disk $\{|z| < \frac{1}{2}|w_n|\}$. If $a_k = E(\frac{z}{w_k}, \lambda_k) - 1$, then, for $|z| < \frac{1}{2}|w_n|$ we have

$$|a_k| \leq 4|\frac{z}{w_k}|^{\lambda_k+1} \leq (\frac{1}{2})^{\lambda_k-1}$$

as soon as $w_k \geq w_n$. It is enough to pick λ_k so that $\prod(1 + (\frac{1}{2})^{\lambda_k-1})$ converges. It is clear that this can be done. In fact, as we see below, we can take $\lambda_k = k$. QED

(12.12) Lemma. $\prod(1 + |a_k|)$ converges if and only if $\sum|a_k|$ converges.

Proof: Assume $\sum|a_k|$ converges. Then

$$\prod_{k=m}^{n} (1 + |a_k|) - 1 = \sum\prod|a_k|$$

$$= \sum_{k=m}^{n} |a_k| + \sum_{m}^{n} |a_{k_1} a_{k_2}| + \cdots$$

$$\leq \sum_{k=m}^{n} |a_k| + \left(\sum_{k=m}^{n} |a_k|\right)^2 + \cdots$$

$$= \frac{\sum_{k=m}^{n} |a_k|}{1 - \sum_{k=m}^{n} |a_k|}$$

Thus $\lim_{m,n\to\infty} \prod_{k=m}^{n} (1 + |a_k|) - 1 = 0$. Conversely, $\sum_{k=m}^{n} |a_k| \leq \prod_{k=m}^{n} (1 + |a_k|) - 1$ so the sum converges if the product does. QED

(12.13) Conclusion of Theorem (12.2): the general case. If $G \neq \mathbb{C}$ and $w \in G$, let w' denote any point in $\mathbb{C}\setminus G$ such that $|w - w'| = \text{dist}(w,\mathbb{C}\setminus G)$. Now we divide W into two disjoint classes A and B. Put w_n into A if $|w_n - w_n'| > \frac{1}{|w_n|}$ and into B otherwise. Then

$$\lim_{\substack{n\to\infty \\ w_n\in A}} w_n = \infty \quad \text{and} \quad \lim_{\substack{n\to\infty \\ w_n\in B}} |w_n - w_n'| = 0.$$

I.e. the sequence A tends to ∞ while B tends to the boundary of G. To see this observe that if infinitely many $w_n \in A$ satisfy $|w_n| < M$, then they also satisfy $|w_n - w_n'| > \frac{1}{M}$ so there would be a limit point w with

$$\text{dist}(w,\mathbb{C}\setminus G) > \frac{1}{M}$$

i.e. $w \in G$. This contradicts the assumption of admissibility. If the sequence B does not tend to ∂G then $\frac{1}{|w_n|} \geq |w_n - w_n'| \geq \delta > 0$ for infinitely many n, i.e. $|w_n| \leq 1/\delta$. Any limit point w of these w_n must satisfy $\text{dist}(w,\mathbb{C}\setminus G) \geq \delta$; again we contradict admissibility.

Now define f_1 to be an entire function with $Z(f_1) = A$, using the construction in the first part (case $G = \mathbb{C}$). Define f_2 by

$$f_2(z) = \prod_{w_n\in B} \left(1 - \left(\frac{w_n - w_n'}{z - w_n'}\right)^{\rho_n}\right)$$

where ρ_n are positive integers chosen to make the product absolutely convergent in $H(G)$. This is possible since, for any compact set $K \subseteq G$, $\frac{w_n - w_n'}{z - w_n'} \to 0$ uniformly in K. This means $\sum \left|\frac{w_n - w_n'}{z - w_n'}\right|^{\rho_n} < +\infty$ for a suitable sequence ρ_n ($\rho_n = n$ will do). Finally, if we let $f(z) = f_1(z)f_2(z)$ then $Z(f)$ will be $A \cup B$. QED

The theorem just finished says that a function in $H(G)$ can be constructed which has any given admissible sequence as its zero set. It is almost a general principle that what can be done with zeros can be done with any sequence of values. Thus we have the following.

(12.14) Germay Interpolation Theorem. Given any admissible sequence $W = \{w_n\}$ in G with multiplicity 1 and any sequence $S = \{s_n\}$ of complex numbers, there is a function $f \in H(G)$ such that

$f(w_n) = s_n$ for $n = 1,2,3,\ldots$.

Proof: Choose a function $g \in H(G)$ which has a simple zero at each point of W. Let $W = A \cup B$ as in the previous proof so that, again,

$$\lim_{\substack{n\to\infty \\ w_n \in A}} |w_n| = \infty \quad \text{and} \quad \lim_{\substack{n\to\infty \\ w_n \in B}} |w_n - w_n'| = 0.$$

For suitable non-negative integers ρ_n define

$$f_n(z) = \begin{cases} s_n \dfrac{g(z)}{z - w_n} \dfrac{1}{g'(w_n)} \left(\dfrac{z}{w_n}\right)^{\rho}, & w_n \in A \\[2ex] s_n \dfrac{g(z)}{z - w_n} \dfrac{1}{g'(w_n)} \left(\dfrac{w_n - w_n'}{z - w_n'}\right)^{\rho_n}, & w_n \in B. \end{cases}$$

Notice that $f_n(w_n) = s_n$, and $f_n(w_k) = 0$ if $k \neq n$. By choosing the numbers ρ_n large enough, the series $\sum f_n$ converges in $H(G)$ to a function with the required properties. (More precisely, let $K_1 \subseteq K_2 \subseteq \cdots$ be an exhaustion of G by compact sets, and let $M_{n,k} = \max\{|f_n(z)| : z \in K_k\}$. Clearly exponents $\rho_{n,k}$ can be chosen so that $\sum_n M_{n,k}$ converges. This makes $\sum f_n(z)$ converge uniformly on K_k. Choosing $\rho_n = \max\{\rho_{n,k} : k = 1,2,\ldots,n\}$ causes $\sum f_n(z)$ to converge uniformly on all K_k.) QED

This interpolation theorem can be extended to cover arbitrary multiplicities: If $W = \{w_n\}$ is a sequence and w_n has multiplicity m_n then $f \in H(G)$ can be constructed which takes the value s_n at w_n with multiplicity m_n. In fact, a more general theorem can be proved to the effect that the first m_n derivatives of f can be specified at each w_n. To prove this we need some material on meromorphic functions.

(12.15) Definition. Let G be an open set in $\mathbb{C}$. A function f is said to be meromorphic in G if there exists an admissible sequence W in G such that

(i) $f \in H(G \setminus W)$, and

(ii) f has a pole at each point in W.

We do not exclude poles of order zero, so that elements of $H(G)$ are meromorphic in G. If f is meromorphic in G and w is one of its poles, then f has an expansion $f(z) = \sum_{n=-m}^{\infty} c_n(z - w)^n$, with $c_{-m} \neq 0$, which converges in some deleted disk about w.

(12.16) Definition. The rational function $\sum_{n=-m}^{-1} c_n(z - w)^n$ is called the principal part of f at w.

The next theorem says that meromorphic functions can be constructed to have arbitrarily specified principal parts at arbitrarily specified poles.

(12.17) Mittag-Leffler Theorem. Let G be an open set in $\mathbb{C}$ and $W = \{w_n : n = 1,2,\ldots\}$ an admissible sequence (multiplicity one) in G. Let $\{P_n\}$ be a sequence of polynomials without constant terms. Then there exists a function $f \in H(G\setminus W)$ with poles on W such that the principal part of f at w_n is $P_n(\frac{1}{z - w_n})$.

Proof: Again let $W = A \cup B$ as in the proof of (12.2) so that

$$\lim_{w_n \in A} |w_n| = +\infty \quad \text{and} \quad \lim_{w_n \in B} |w_n - w_n'| = 0$$

where w_n' is a point on ∂G with $|w_n - w_n'| = \text{dist}(w_n, G)$. We will construct two functions, f_1 and f_2, where f_1 has poles at $w_n \in B$ and f_2 has poles at $w_n \in A$, each with the appropriate principal parts. Then $f = f_1 + f_2$ will be the required function.

For each $w_n \in B$, let $F_n = \{z \in G : |z - w_n'| > 2|w_n - w_n'|\}$. Note that

$$\frac{1}{z - w_n} = \sum_{k=0}^{\infty} \frac{(w_n - w_n')^k}{(z - w_n')^{k+1}}$$

and the convergence is uniform for $z \in F_n$. Let

$$g_n(z) = P_n\left(\frac{1}{z - w_n}\right) - P_n\left(\sum_{k=0}^{\rho_n} \frac{(w_n - w_n')^k}{(z - w_n')^{k+1}}\right)$$

where ρ_n is chosen so that $|g_n(z)| < 2^{-n}$ for $z \in F_n$. Let $f_1(z) = \sum_{w_n \in B} g_n(z)$. If $z_0 \in G$ then, for large enough n_0, $z_0 \in \text{int } F_n$ for all $n > n_0$. Consequently

$$\sum_{\substack{n>n_0 \\ w_n\in B}} g_n(z)$$

converges uniformly in a neighborhood of z_0 to a holomorphic function and

$$\sum_{\substack{n\leq n_0 \\ w_n\in B}} g_n(z)$$

is a rational function with poles among $w_1, w_2, \ldots, w_{n_0}$ and principal parts $P_n(\frac{1}{z - w_n})$, $n \leq n_0$, $w_n \in B$. Thus $\sum_{w_n\in B} g_n(z)$ is meromorphic in G with principal parts $P_n(\frac{1}{z - w_n})$, $n \in B$.

To construct f_2, let $G_n = \{z \in G : |z| < |w_n|/2\}$ for $w_n \in A$. Observe that

$$\frac{1}{z - w_n} = - \sum_{k=0}^{\infty} \frac{z^k}{w_n^{k+1}}$$

with uniform convergence on G_n. Let

$$h_n(z) = P_n(\frac{1}{z - w_n}) - P_n\left(- \sum_{k=0}^{\sigma_n} \frac{z^k}{w_n^{k+1}}\right)$$

where σ_n is chosen so that $|h_n(z)| < 2^{-n}$ for $z \in G_n$. Finally, letting $f_2(z) = \sum_{w_n\in A} h_n(z)$, it follows as before that f_2 is meromorphic in G with poles at $w_n \in A$ and principal parts $P_n(\frac{1}{z - w_n})$. Clearly $f = f_1 + f_2$ has the required properties. QED

The Mittag-Leffler Theorem gives us the tool we need to prove the generalization of the interpolation theorem.

(12.18) <u>Extended Interpolation Theorem</u>. <u>Let</u> W <u>be an admissible sequence in</u> G <u>such that</u> $w_n \in W$ <u>has multiplicity</u> m_n. <u>Let</u> $s_{n,k}$, $n = 1,2,\ldots; k = 0,1,\ldots,m_n - 1$ <u>be complex numbers</u>. <u>Then there is a function</u> $f \in H(G)$ <u>such that</u>

$$\frac{1}{k!} f^{(k)}(w_n) = s_{n,k}, \; n = 1,2,\ldots; k = 0,1,\ldots,m_n - 1.$$

Proof: We need to prove f can be constructed so its Taylor series at w_n has the form $s_{n,0} + s_{n,1}(z - w_n) + \cdots + s_{n,m_n-1}(z - w_n)^{m_n-1} + \cdots$. Let g be any function in $H(G)$ which has a zero of mulitplicity m_n at each w_n. Let $P_n(z) = a_{n,1}z + a_{n,2}z^2 + \quad + a_{n,m_n}z^{m_n}$ be a polynomial with coefficients $a_{n,k}$ chosen so that $P_n(\frac{1}{z - w_n})g(z)$ has the proper Taylor series at w_n. Assuming this can be done, construct h by the Mittag-Leffler Theorem with principal parts $P_n(\frac{1}{z - w_n})$ at w_n, and let $f(z) = h(z)g(z)$.

We need to show that P_n can be constructed. We assume for simplicity that $w_n = 0$ and we omit the subscript n. We have to select $a_0,\ldots,a_{m-1}$ so that

$$\left(\frac{a_1}{z} + \frac{a_2}{z^2} + \cdots + \frac{a_m}{z^m}\right)(b_m z^m + b_{m+1}z^{m+1} + \cdots)$$

$$= s_0 + s_1 z + \cdots + s_{m-1}z^{m-1} + \cdots$$

where $b_m, b_{m+1},\ldots$ are the coefficients of $g(z)$ and $b_m \neq 0$. This gives equations

$$a_m b_m = s_0$$

$$a_{m-1}b_m + a_m b_{m-1} = s_1$$

$$\vdots$$

$$a_1 b_m + \cdots + a_m b_{2m-1} = s_{m-1}.$$

It is clear that these equations can be solved in the order they appear, obtaining first $a_m = s_0/b_m$, $a_{m-1} = (s_1 - a_m b_{m-1})/b_m$, etc.

NOTES: In this chapter we have intentionally avoided abstract methods. Thus we have not used Runge's Theorem because our proof of it uses the duality theorem. The Mittag-Leffler Theorem could be simplified by using it. In fact, the proof we present uses one of the arguments occurring in a more constructive proof of Runge's Theorem. In a later chapter, we give a functional analysis proof of the theorems of this chapter. The reader will note the typical contrast between constructive and abstract methods: constructive methods are stronger in that explicit formulas and estimates are obtained while abstract mehtods are stronger in that more general theorems usually result and interesting connections with apparently unrelated problems are often exposed.

The different points of view often lead to different lines of investigation.

Exercises

1. If $a_k \geq 0$ for $k = 1,2,3,\ldots$ show that $\prod(1 + a_k)$ is convergent if and only if the partial products $\prod_{k=1}^{n}(1 + a_k)$ form a bounded sequence. Prove Proposition 12.8.

2. Show that $z\prod_{n=1}(1 - \frac{z^2}{n^2})$ is an entire function with zeros precisely at the integers $0, \pm 1, \pm 2, \ldots$. What function might this be?

3. If $\prod(1 + a_k)$ is absolutely convergent and $b_k = a_{\sigma(k)}$, where $\sigma : \{1,2,\ldots\} \to \{1,2,\ldots\}$ is a permutation, show that

$$\prod(1 + b_k) = \prod(1 + a_k).$$

4. Does there exist a non-zero entire function which has at least one zero in each wedge? i.e. in $\{z = re^{i\theta} : r > 0, \theta_0 < \theta < \theta_0 + \alpha\}$ for any choice of θ_0 and $\alpha > 0$?

5. Does there exist a non-zero entire function that has at least one zero in every convex set of area ≥ 1?
Hint: It might help to know that every convex set C contains a triangle T with area $T \geq \frac{1}{4}$ area C. Proof: Take two points of C at maximum distance apart and take a third point of C at maximum distance from the line through the first two. Then the triangle determined by these three points works:

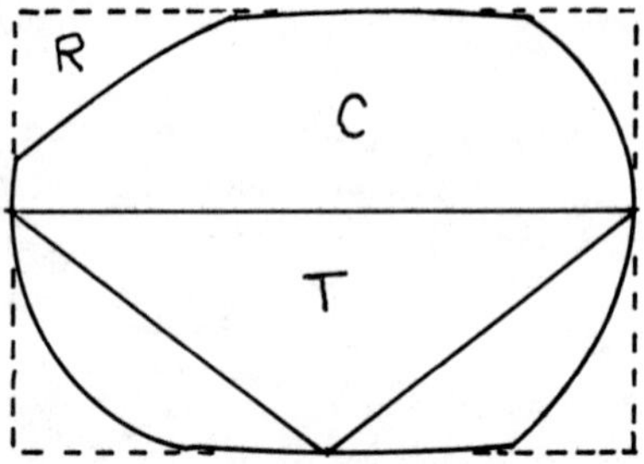

Figure 12.1

6. Let u_k be a sequence of bounded analytic functions in a domain G such that $|u_k(z)| \leq 1$ for all k and all $z \in G$. In addition, let there exist a point $z_0 \in G$ such that $\forall k$, $u_k(z_0) = |u_k(z_0)|$. Show that $\prod u_k$ converges in $H(G)$ if and only if $\sum 1 - |u_k(z_0)| < +\infty$.

7. Show that every meromorphic function in G is a quotient of holomorphic functions.

8. a) Given an entire function f let $f^{\#}(z) = \sum_{n=1} f(1/n)z^n$. Does there exist a non-zero entire function with $f = f^{\#}$?

*b) Same question with $f^*(z) = \sum_{n=0} f(n)z^n$.

*c) Same as a) except we require $f \in H(\mathbb{D})$.

9. For entire functions write $f \sim g$ to mean that $\lim_{n\to\infty} f(z_n)/g(z_n)$ exists and is not zero for every sequence of points $z_n \to \infty$. Show that $f \sim g$ if and only if there exist polynomials P and Q of equal degree and an entire function H such that $f = PH$ and $g = QH$.

10. Show that for connected open $G \subseteq \mathbb{C}$ there is a function $f \in H(G)$ such that f cannot be extended to a function analytic in any larger connected open set.

§13. Ideals in $H(G)$

We use the results of the previous section to derive some descriptive results on ideals of holomorphic functions.

(13.1) Definition. In a commutative ring R, an ideal is a subset I of R such that $a,b \in I$ implies $a - b \in I$ and if $r \in R, a \in I$ then $ra \in I$. Given $a_1, a_2, \ldots, a_n \in R$ the set $I = \{\rho_1 a_1 + \rho_2 a_2 + \cdots + \rho_n a_n : \rho_j \in R, j = 1,2,\ldots,n\}$ is called the ideal generated by $\{a_j : j = 1,2,\ldots,n\}$ and is commonly denoted $I = (a_1,\ldots,a_n)$. I is said to be finitely generated.

It is easy to verify that $(a_1,\ldots,a_n)$ is indeed an ideal. Strictly speaking, we should be saying left ideal everywhere but, as we shall be dealing only with commutative rings, there will be no distinction among left, right and two-sided ideals.

(13.2) If $a,b \in R$, we say a divides b and write $a|b$ if there is an element $c \in R$ with $ac = b$. If R has a multiplicative identity (an element 1 such that $1r = r$ for all $r \in R$) then $u \in R$ is called a unit if $u|1$. We also say such a unit u is invertible. In $H(G)$ the units are those functions without zeros. More generally, in $H(G)$, $f|g$ if and only if $Z(f) \subseteq Z(g)$ counting multiplicity. A function $\varphi \in H(G)$ is a common divisor of a collection F of functions in $H(G)$ provided that $\varphi|f$ for each $f \in F$. If φ has the property that any other common divisor ψ satisfies $\psi|\varphi$, then φ is called a greatest common divisor and we write $\varphi = \gcd(F)$.

We now suppose G is connected and $0 \notin F$. Let $Z(F) = \bigcap_{f \in F} Z(f)$, where we use the convention that a point $z \in Z(F)$ has multiplicity equal to the minimum of the multiplicities it has in the various $Z(f)$'s. Then φ is a common divisor of F if and only if $Z(\varphi) \subseteq Z(F)$ counting multiplicities.

(13.3) Proposition. Any collection $F \subseteq H(G)$ with $0 \notin F$ has a greatest common divisor. If φ and ψ are both greatest common divisors of F, then $\psi = u\varphi$ where u is a unit in $H(G)$.

Proof: Apply Theorem (12.2) to construct a function $\varphi \in H(G)$ which has zero set equal to $Z(F)$ ($Z(F)$ is admissible since it is contained in the admissible sequences $Z(f)$.) If ψ is a common divisor of F then $Z(\psi) \subseteq Z(F) = Z(\varphi)$ so $\psi|\varphi$. Thus $\varphi = \gcd F$. If ψ is any other gcd of F then by definition $\psi|\varphi$ and $\varphi|\psi$ i.e. $\psi = u\varphi$ and $\varphi = v\psi$ where $u,v \in H(G)$. Clearly $uv = 1$ so u is a unit. QED

Let $M(G)$ denote the class of meromorphic functions in G.

(13.4) Proposition. If $f \in M(G)$ then there exist $g,h \in H(G)$ such that $f = g/h$.

Proof: An easy consequence of the Extended Interpolation Theorem (12.18) (see Exercise 7, Chapter 12): Construct h so that $hf \in H(G)$. QED

(13.5) Definition. An ideal $I \in R$ is called principal if it has one generator, i.e. $I = (a)$ for some $a \in R$.

(13.6) Helmer's Theorem. In $H(G)$, every finitely generated ideal is principal.

Proof: It is enough to prove the theorem for an ideal with two generators. We prove $(f,g) = (h)$ where h is the greatest common divisor of f and g. Write $f = f_1h$ and $g = g_1h$. Then f_1 and g_1 have no common zeros. (If z_0 is a zero of both f_1 and g_1 then $(z - z_0)h(z)$ would be a common divisor of f and g.) It is easy to see that $(f,g) = (f_1,g_1)\cdot(h)$, so we need only prove $(f_1,g_1) = (1) = H(G)$. Thus, we seek $a,b \in H(G)$ such that $af + bg = 1$. This will be the case if we can construct a function b with $(1 - bg)/f \in H(G)$. Therefore choose $b(z) \in H(G)$, according to the Extended Interpolation Theorem, so that for each $w \in Z(f)$ with multiplicity $m = m(w)$, $1 - bg$ has a zero at w of multiplicity m. This amounts to a soluable system of linear equations in finitely many derivatives of b at $z = w$. Then $Z(1 - bg) \supseteq Z(f)$ with multiplicity, so $(1 - bg)/f \in H(G)$ and Helmer's Theorem is proved. QED

As an application of Helmer's Theorem we can characterize the closed ideals in $H(G)$.

(13.7) Theorem. If G is a region, the closed ideals in $H(G)$ are precisely the principal ideals.

Proof: First we show that principal ideals are closed. If $f = 0$, then $(f) = \{0\}$ which is certainly closed. If $f \neq 0$ and if $g_n f \to h$, we have $Z(h) \supseteq Z(f)$ so $h/f \in H(G)$, i.e. $h = (\frac{h}{f}) f \in (f)$. This proves (f) is closed. (Note: In concluding $Z(h) \supseteq Z(f)$ we are using the continuity of point evaluation and the operation of taking derivatives, since $Z(f)$ and $Z(h)$ are sets with multiplicity.)

Now suppose I is a closed ideal. Let $\varphi = \gcd(I)$ and consider $I' = \{f/\varphi : f \in I\}$. We prove that $I' = (1)$. Because I is closed, I' is closed. For if $f_n/\varphi \to g$ in $H(G)$, then $f_n \to \varphi g$, hence $\varphi g \in I$ and $g \in I'$. Since $I = \varphi I'$ it is enough to prove $I' = H(G)$. The crucial property of I' is that its zero set is empty: $Z(I') = \bigcap_{f \in I'} Z(f) = \emptyset$. Choose compact sets K, K' in G with $K \subseteq \operatorname{int} K'$, and let $K'' = (K')^{G}$ (the hull of K' with respect to G, Definition (10.3)). For each $z_0 \in K'$ there is an $f_0 \in I'$ such that $f_0(z_0) \neq 0$. By continuity, f_0 does not vanish in some neighborhood of z_0. By the Heine-Borel Theorem, there are finitely many functions $f_1, f_2, \ldots, f_n \in I'$ that have no common zeros on K''. By Helmer's Theorem there is a function $h_K \in I'$ such that $(f_1, f_2, \ldots, f_n) = (h_K)$. Clearly h_K can have no zeros on K''. By Runge's Theorem, for any $\varepsilon > 0$, there is a rational function R_ε with poles in $\mathbb{C}^\wedge \setminus G$ such that $\| R_\varepsilon - \frac{1}{h_K} \|_K < \varepsilon$ and so $\| R_\varepsilon h_K - 1 \| < \varepsilon \| h_K \|_K$. But $R_\varepsilon h_K \in I'$ since I' is an ideal and what we have just shown is that the function 1 can be approximated uniformly on compact sets by elements of I'. Since I' is closed in $H(G)$, $1 \in I'$ and the result is proved. QED

(13.8) Definition. In a commutative ring R, a maximal ideal is a proper ideal which is not properly contained in any other proper ideal. That is, I is maximal if whenever $I \subseteq J$ and J is an ideal then $J = I$ or $J = R$.

If R has a unit, 1, then 1 is not contained in any proper ideal. The Hausdorff Maximality Principle, applied to the ideals in R ordered by $\subseteq$, shows that in rings with unit, every proper ideal is contained in a maximal ideal.

(13.9) Proposition. The closed maximal ideals in $H(G)$ are precisely the ideals of functions that vanish at a single point.

Proof: If $z_0 \in G$ and $I_{z_0} = \{f \in H(G) : f(z_0) = 0\}$, then I_{z_0} is just the principal ideal generated by $z - z_0$. It is clear that

I_{z_0} is maximal. Conversely if I is closed and maximal then $I = (f)$ for some $f \in H(G)$. Clearly $Z(f) \neq \emptyset$ or we would have $I = H(G)$. If $z_0 \in Z(f)$ then $(z - z_0)|f$, which implies $I \subseteq I_{z_0}$. Since I is maximal and $I_{z_0} \neq H(G)$, we must have $I = I_{z_0}$. QED

(13.10) Definition. An ultrafilter $\mathcal{U}$ on a set S is a non-empty collection of subsets of S with the following properties

(1) $\emptyset \notin \mathcal{U}$,

(2) If $U, V \in \mathcal{U}$ then $U \cap V \in \mathcal{U}$

(3) Any superset of a set in $\mathcal{U}$ belongs to $\mathcal{U}$.

(4) $\mathcal{U}$ is maximal with respect to these properties.

The meaning of (4) is the following: If $\mathcal{U}'$ is a collection of subsets of S with properties (1), (2) and (3), and if $\mathcal{U} \subseteq \mathcal{U}'$, then $\mathcal{U} = \mathcal{U}'$. Notice that (1) and (3) imply $S \in \mathcal{U}$. Property (4) may be replaced by the following.

(4') If A, B are any subsets of S with $A \cup B \in \mathcal{U}$, then either $A \in \mathcal{U}$ or $B \in \mathcal{U}$.

Proof: Let $\mathcal{U}$ satisfy (1)-(4) and suppose $A \cup B \in \mathcal{U}$, $B \notin \mathcal{U}$. Now $(A \cup B) \cap U$ is a non-empty member of $\mathcal{U}$ for every $U \in \mathcal{U}$. Then $A \cap U$ must be non-empty (else B, as a superset of $(A \cup B) \cap U$, would belong to $\mathcal{U}$, contrary to assumption) for all $U \in \mathcal{U}$ and it is easy to see that $\mathcal{U} \cup \{A \cap U : U \in \mathcal{U}\}$ satisfies (1)-(3) and contains A. By maximality $A \in \mathcal{U}$.

Conversely, let (1)-(3) and (4') be satisfied. Suppose $\mathcal{U} \subseteq \mathcal{U}'$ where $\mathcal{U}'$ satisfies (1)-(3). Let $A \in \mathcal{U}'$. Then, by (4'), either A or $S \setminus A$ belongs to $\mathcal{U}$. But if $S \setminus A \in \mathcal{U}$ then $S \setminus A \in \mathcal{U}'$ so $\emptyset = (S \setminus A) \cap A \in \mathcal{U}'$ contrary to (1). Thus $S \setminus A \notin \mathcal{U}$, so $A \in \mathcal{U}$. Since A is arbitrary $\mathcal{U}' = \mathcal{U}$, and (4) is proved.

(13.11) A non-empty collection satisfying only (1)-(3) is called a filter, An example of an ultrafilter is the collection of all subsets of S containing a fixed point s_0. Such an ultrafilter is termed principal. If $\mathcal{U}$ is a non-principal ultrafilter then

(5) $\bigcap_{U \in \mathcal{U}} U = \emptyset$

and conversely. To see this, suppose $s \in \bigcap_{U \in \mathcal{U}} U$. Then the collection of subsets which contain s, contains $\mathcal{U}$. By maximality, $\mathcal{U}$ is principal. Thus (5) holds for non-principal ultrafilters.

(13.12) Proposition. *If S is infinite then there is a non-prinicipal ultrafilter on S.*

Proof: Let $\mathcal{B}$ consist of all subsets of S whose complements are finite (the cofinite sets). Observe that (1)-(3) and (5) are satisfied. Consider all filters F that contain $\mathcal{B}$ and order them by inclusion. The union of any chain of filters is again a filter. Let $\mathcal{U}$ be the union of any maximal chain of filters containing $\mathcal{B}$ (whose existence is quaranteed by the Hausdorff Maximality Principle), then $\mathcal{U}$ will satisfy (4) and (5). QED

(13.13) Proposition. *If G is a region, then $H(G)$ contains non-closed maximal ideals.*

Proof: Let Ω be an admissible sequence (of multiplicity 1) in G, and let $\mathcal{U}$ be a non-principal ultrafilter on Ω. Let I consist of all $f \in H(G)$ such that $f(U) = 0$ for some $U \in \mathcal{U}$. Clearly, $I \neq \emptyset$ and $I \neq (1)$. We prove I is maximal. Suppose $f \notin I$. Then $Z(f) \cap \Omega \notin \mathcal{U}$ so that if $U = \Omega \setminus Z(f)$, then $U \in \mathcal{U}$. Construct $g \in H(G)$ such that $Z(g) = U$, and hence $g \in I$. Since $Z(f) \cap Z(g) = \emptyset$, we have $(f,g) = (1)$. It follows that I is maximal, because if $J \supseteq I$, $J \neq I$, then J contains a function $f \notin I$. The above construction yields $g \in I$ with $J \supseteq (f,g) = H(G)$. Moreover, I is not closed since it is not principal: $Z(I) = \bigcap_{U \in \mathcal{U}} U = \emptyset$. QED

So far the ring $H(G)$ shares many properties with $\mathbb{Z}$, the ring of all integers. They are both integral domains ($fg = 0 \Rightarrow f = 0$ or $g = 0$) such that each finitely generated ideal is principal. However, in $\mathbb{Z}$, every prime ideal is maximal. Is this the case in $H(G)$?

(13.14) Definition. *An ideal $I \neq 0$ is prime if $fg \in I$ implies that either $f \in I$ or $g \in I$.*

(13.15) Proposition. *If G is a region, then there is a prime ideal in $H(G)$ that is not maximal.*

Proof: Let $W = \{w_m\}$ be an admissible sequence in G, with multiplicity 1. for each $F \in H(G)$, let φ_f be defined by $\varphi_f(n)$ is the multiplicity of the zero of f at w_n. ($\varphi_f(n) = 0$ means that $f(w_n) \neq 0$.) Now let $\mathcal{U}$ be a non-principal ultrafilter on the positive integers. Let Φ be the class of functions $\varphi : \{1,2,3,\ldots\} \to \{\infty,0,1,2,\ldots\}$ such that for each integer k, $E_k(\varphi) \in \mathcal{U}$, where $E_k(\varphi) = \{n : \varphi(n) \geq k\}$. Then Φ is non-empty since the function φ_0 defined by $\varphi_0(n) = n$ belongs to Φ. Let I be the set of functions $f \in H(G)$ such that $\varphi_f \in \Phi$. We shall show that I is a prime ideal that is not maximal. First, Φ contains any function φ such that $\lim_{n\to\infty} \varphi(n) = \infty$. This is because $\mathcal{U}$ contains all co-finite sets (since it contains no finite ones) and $E_k(\varphi)$ is cofinite for such a φ. Thus I contains non-zero functions. To see that I is an ideal, let f, $g \in I$. Then $\varphi_{f-g} \geq \min(\varphi_f,\varphi_g)$ so that

$$E_k(\varphi_{f-g}) \geq \{n : \min(\varphi_f(n),\varphi_g(n)) \geq k\}$$

$$= \{n : \varphi_f(n) \geq k\} \cap \{n : \varphi_g(n) \geq k\}$$

$$= E_k(\varphi_f) \cap E_k(\varphi_g) \in \mathcal{U}.$$

Therefore $E_k(\varphi_{f-g}) \in \mathcal{U}$, proving $f - g \in I$. Also if $f \in I$ and $g \in H(G)$, then $fg \in I$ because $\varphi_{fg} \geq \varphi_f$ so that $E_k(\varphi_{fg}) \supseteq E_k(\varphi_f)$ which implies $E_k(\varphi_{fg}) \in \mathcal{U}$. Now suppose $fg \in I$. Then

$$E_{2k}(\varphi_{fg}) = E_{2k}(\varphi_f + \varphi_g) = \bigcup_{j=0} [E_j(\varphi_f) \cap E_{2k-j}(\varphi_g)]$$

since $\varphi_f(n) + \varphi_g(n) \geq 2k$ if and only if $\varphi_f(n) \geq 2k - \varphi_g(n)$. By property (4') of ultrafilters, there is some j with $0 \leq j \leq 2k$ and

$$E_j(\varphi_f) \cap E_{2k-j}(\varphi_g) \in \mathcal{U}.$$

Since either $j \geq k$ or $2k - j \geq k$, we have either

$$E_k(\varphi_f) \supseteq E_j(\varphi_f) \cap E_{2k-j}(\varphi_g) \in \mathcal{U}.$$

or

$$E_k(\varphi_g) \supseteq E_j(\varphi_f) \cap E_{2k-j}(\varphi_g) \in \mathcal{U}.$$

Thus, either $E_k(\varphi_f) \in \mathcal{U}$ or $E_k(\varphi_g) \in \mathcal{U}$. It follows that either $E_k(\varphi_f) \in \mathcal{U}$ for infinitely many k or $E_k(\varphi_g) \in \mathcal{U}$ for infinitely many k. In the first case, $E_k(\varphi_f) \in \mathcal{U}$ for all k (since $E_k(\varphi) \supseteq E_m(\varphi)$ if $m > k$) and so $f \in I$. In the second case, of course, $g \in I$.

Finally, I is not maximal. For we can take $F \in H(G)$ with a simple zero at each w_n. Then $F \notin I$ because $E_2(\varphi_F) = \emptyset \notin \mathcal{U}$. Let J be the set of all functions $h \in H(G)$ of the form $f + aF$ for some $f \in I$ and $a \in H(G)$. Clearly $J \supseteq I$ but $J \neq I$ and $J \neq H(G)$ because $1 \notin J$. (Every element of J has a non-empty zero set.) Since J is an ideal, I cannot be maximal. QED

(13.16) If R is a commutative ring and I is an ideal in R then R/I (also written $\frac{R}{I}$) denotes the following ring. For $f, g \in R$ write $f \equiv g (\text{mod } I)$ to mean that $f - g \in I$. (We usually omit the "mod I".) Clearly, $\equiv$ is an equivalence relation which is compatible with the operations in R. This means if $f \equiv g$ and $h \equiv k$, then $f + h \equiv g + k$ and $fh \equiv gk$. We denote by [f] the equivalence class that contains f. Defining $[f] + [g] = [f + g]$ and $[f][g] = [fg]$, we have a ring that is denoted by R/I or $\frac{R}{I}$. There is a natural homomorphism $\varphi : R \to R/I$ given by $\varphi(f) = [f]$ which satisfies $\varphi^{-1}(0) = I$.

(13.17) Proposition. *If R is a commutative ring with unit and I is a maximal ideal, then R/I is a field.*

Proof: It must be shown that there is a unique solution [f] to the equation $[f][h] = [1]$ for any given $[h] \in R/I$, $[h] \neq [0]$. This says that in R, we must find an $f \in R$ satisfying $fh - 1 \in I$ whenever $h \notin I$; and that any two such f must differ by an element of I. But if $fh - 1 \in I$ and $f'h - 1 \in I$, then $(f - f')h \in I$. Since I is maximal, it is prime. (Why?) Therefore, $f - f' \in I$. To show that f exists, let J be the ideal consisting of all elements of the form $fh - \lambda$, with $f \in R$ and $\lambda \in I$. Then $J \supseteq I$, and $h \in J$ so $J \neq I$. Thus $J = R$, since I is maximal; hence $1 \in J$. QED

(13.18) Theorem (Henriksen). *If I is a maximal ideal in $H(G)$ then $H(G)/I$ is algebraically closed*, i.e. *every equation* $a_n x^n + a_{n-1} x^{n-1} + \cdots + a_0 = 0$ *with* $a_k \in H(G)/I$ *has a solution* $x \in H(G)/I$.

Remark. Henriksen actually proved more, namely that $H(G)/I$ is isomorphic to $\mathbb{C}$. But the isomorphism can be a "bad" one in a certain sense.

Proof: If $Z(I) = \bigcap_{f \in I} Z(f) \neq \emptyset$ then for some $z_0 \in G$, $I = \{f : f(z_0) = 0\}$. Define $\psi : H(G)/I \to \mathbb{C}$ by $\psi([f]) = f(z_0)$. It is trivial to verify that ψ is an isomorphism. Since $\mathbb{C}$ is algebraically closed, so is $H(G)/I$.

Next, suppose $Z(I) = \emptyset$ and consider the equation

(a) $$[f_n]X^n + \cdots [f_1]X + [f_0] = 0.$$

where $[f_n] \neq [0]$, i.e. $f_n \notin I$. Now there is a function $f \in I$ such that $Z(f) \cap Z(f_n) = \emptyset$. For otherwise, the ideal J given by $J = \{g + af_n : g \in I, \ a \in H(G)\}$ would be a superideal of I which is not $H(G)$ since every function in J would have a non-empty zero set. Since I is maximal it follows that $I = J$ and $f_n \in I$, a contradiction. Let F be a function with simple zeros at the zeros of f. Then $F \in I$, for otherwise there would exist an ideal $J \neq H(G)$, $J \supseteq I$, with $F \in J$ (as in the proof of 13.15), contradicting the maximality of I. Let $\{z_1, z_2, \ldots\}$ be the zeros of F. Choose numbers $w_1, w_2, \ldots$ so that $f_n(z_j)w_j^n + \cdots + f_1(z_j)w_j + f_0(z_j) = 0$ for $j = 1,2,3,\ldots$. This is possible because $f_n(z_j) \neq 0$. Now construct $g \in H(G)$ with $g(z_j) = w_j$. The function $\lambda = f_n g^n + \cdots + f_1 g + f_0$ has a zero at each z_j. This means $F|\lambda$ and so $\lambda \in I$. Choosing $X = [g]$, we have solved (a). QED

NOTES: Henriksen's work on ideals can be found in [M. Henriksen, 1 and 2]. Helmer's Theorem is from [O. Helmer].

Exercises

1. Show that a non-principal ultrafilter contains no finite sets and all co-finite sets.

2. Show that a maximal ideal is prime.

3. Given a non-constant polynomial $p(z)$, produce a non-constant polynomial $q(z)$ such that $p(z)$ and $q(p(z))$ generate a dense ideal in E.

4. Characterize all closed <u>differential ideals</u> in $H(G)$. An ideal I will be called <u>differential</u> if $f' \in E$ whenever $f \in I$.

5. (a) Let I be a <u>proper</u> differential ideal in $H(G)$ (see Exercise 4). For each $f \in I$ define $\varphi_f(z)$ = order of the zero f has at z. Show that the family $\mathcal{O} = \{\varphi_f : f \in I\}$ has the following properties.

(i) If $\varphi \in \mathcal{O}$ then φ is unbounded

(ii) If $\varphi, \psi \in \mathcal{O}$ then $\min\{\varphi,\psi\} \in \mathcal{O}$

(iii) If $\varphi \in \mathcal{O}$ then $\max\{0,\varphi - n\} \in \mathcal{O}$ for each $n = 1,2,\ldots$.

(iv) if $\varphi \in \mathcal{O}$ then $\{z \in G : \varphi(z) \neq 0\}$ is an admissible sequence.

(b) Prove that if $\mathcal{O}$ satisfies (i) to (iv) then $I = \{f \in H(G) : \exists\psi \in \mathcal{O}$ such that $\varphi_f(z) \geq \psi(z)$ on $G\}$ is a proper differential ideal.

(c) Find an easier way to produce a proper differential ideal.

*(d) Can a maximal ideal be differential?

6. We know that given $f,g \in E$ with no common zeros, there exist $a, b \in E$ such that $af + bg = 1$.

a) Can we choose a to be zero-free?

b) Can we choose both a and b to be zero-free?

*7. Same as problem 6 except E is replaced by $H(\mathbb{D})$ or $H(G)$.

8. Given a sequence $z_n \to \infty$ construct an entire function such that the ideal generated by $\{f(z + z_1), f(z + z_2), \ldots\}$ is not dense in E.

9. Define an ideal I in $H(G)$ to be radical if $f^n \in I$ implies $f \in I$ for any $n = 1,2,\ldots$. Determine the closed radical ideals in $H(G)$.

*10. Fix a positive integer k. Does there exist an infinite set in E such that no subset of k elements generates a dense ideal but every subset of $k + 1$ elements does? This is easy if $k = 1$. Say $k = 2$ or more.

11. Prove Proposition 13.13 by constructing an ideal with no common zeros and invoking the fact that every proper ideal is contained in a maximal one.

§14. The Riemann Mapping Theorem

The Riemann Mapping Theorem implies that, as far as $H(G)$ can tell, all simply connected regions are the "same". To clarify what this means we need the following notion of equivalence.

(14.1) Definition. We say that two regions G and G' are conformally equivalent if there exists a function φ which is holomorphic and univalent in G such that $\varphi(G) = G'$. We write $G \cong G'$ and call φ a conformal mapping.

It is easy to see that this is indeed an equivalence relation. We have seen (5.4) that $G \cong G'$ if and only if $H(G)$ is algebra isomorphic to $H(G')$.

(14.2) The Riemann Mapping Theorem. If G is a simply connected region in $\mathbb{C}$ and $G \neq \mathbb{C}$, then $G \cong \mathbb{D}$, the unit disk. Let $a \in G$. There is a unique conformal mapping function φ such that $\varphi(a) = 0$ and $\varphi'(a) > 0$. We call this the normalized mapping function.

An alternate version of this theorem says that, on the Riemann sphere, there are only three different simply connected regions modulo conformal equivalence. They are (i) the sphere, (ii) the sphere minus the north pole, (iii) the southern hemisphere.

(14.3) Schwarz's Lemma. If $f \in H(\mathbb{D})$, $f(0) = 0$ and $|f(z)| \leq 1$ for all $z \in \mathbb{D}$, then unless $f(z) = e^{i\lambda}z$ for some real number λ and all $z \in \mathbb{D}$, it follows that $|f(z)| < |z|$ for all $z \in \mathbb{D}$, and that $|f'(0)| < 1$.

Proof: Define $g(z) = f(z)/z$ with $g(0) = f'(0)$. Then the maximum modulus theorem says $|g(z)| \leq 1$ and if $|g(z_0)| = 1$ for any $z_0 \in \mathbb{D}$ then g is constant. This is equivalent to the stated result.
QED

(14.4) The function $\varphi(z) = \dfrac{az + b}{cz + d}$ with $ad - bc \neq 0$, are called Möbius transformations (also called linear fractional, fractional linear, linear, bilinear or homographic transformations). Möbius transformations are one-to-one maps of the Riemann sphere onto itself, provided the following conventions are used: If $c \neq 0$ then $\varphi(\infty) = a/c$ and $\varphi(-\frac{d}{c}) = \infty$. If $c = 0$, then φ is a linear polynomial and $\varphi(\infty) = \infty$.

It is left as an exercise to show that any conformal map of the Riemann sphere onto itself must be a Möbius transformation. The Möbius transformations map circles onto circles (on the sphere--in the plane a circle through ∞ becomes a straight line) since each Möbius transformation is a composition of the simpler ones (i) $z + a$ (translation), (ii) $1/z$ (inversion, (iii) $e^{i\lambda}z$, λ real (rotation), (iv) Az, $A > 0$ (dilation). Each of these simpler ones takes circles to circles.

If $z_0 \in \mathbb{D}$ and λ is a real constant, then the Möbius transformation $B(z : z_0, \lambda)$, defined by

$$B(z : z_0, \lambda) = e^{i\lambda}\frac{z - z_0}{1 - \bar{z}_0 z}$$

is called a Blaschke factor. Usually Blaschke factors are normalized by taking $e^{i\lambda}$ to be $+1$, -1 or $-|z_0|/z_0$. The last choice makes $B(0 : z_0, \lambda) > 0$; the first makes $B'(0 : z_0, \lambda) > 0$.

(14.5) Proposition. The most general one-to-one conformal map of $\mathbb{D}$ onto $\mathbb{D}$ is a Blaschke factor.

Proof: First, it is easy to see that $B(z) = B(z : z_0, \lambda)$ takes $\mathbb{D}$ one-to-one onto $\mathbb{D}$, for each Möbius transformation is one-to-one with a one-to-one inverse. Since the inverse of a Blaschke factor is easily computed to be a Blaschke factor, it is sufficient to prove that a Blaschke factor takes $\mathbb{D}$ into $\mathbb{D}$. By the maximum modulus theorem it is enough to show that $B(\partial\mathbb{D}) \subseteq \partial\mathbb{D}$. But

$$\left|\frac{e^{i\theta} - z_0}{1 - e^{i\theta}\bar{z}_0}\right| = \left|e^{-i\theta}\frac{e^{i\theta} - z_0}{e^{-i\theta} - \bar{z}_0}\right| = 1,$$

since $|e^{-i\theta} - \bar{z}_0| = |\overline{e^{i\theta} - z_0}| = |e^{i\theta} - z_0|$.

Next, suppose f is a conformal one-to-one map of $\mathbb{D}$ onto $\mathbb{D}$. If $z_0 = f(0)$ and $g = B(f(z) : z_0, 0)$, then g is a one-to-one map of $\mathbb{D}$ onto $\mathbb{D}$ and $g(0) = 0$. Applying Schwarz's Lemma, we see that $|g(z)| \le |z|$. But applying it to g^{-1} we see that $|g^{-1}(z)| \le |z|$ and hence $|g(z)| \ge |z|$. Consequently $|g(z)| = |z|$ so $g(z) = e^{i\lambda}z$ for some real λ. But $f = B^{-1} \circ g$ where $B(z) = \dfrac{z - z_0}{1 - \bar{z}_0 z}$, and so

$$f(z) = \frac{e^{i\lambda}z - z_0}{1 - \bar{z}_0 e^{i\lambda}z} = B(z : z_0 e^{-i\lambda}, \lambda). \quad \text{QED}$$

The uniqueness part of the Riemann mapping theorem can be proved along similar lines. For, if φ_1 and φ_2 were two normalized mapping functions then $h = \varphi_2 \circ \varphi_1^{-1}$ would be a conformal one-to-one map of $\mathbb{D}$ onto $\mathbb{D}$ with $h(0) = 0$ and $h'(0) > 0$. It follows from 14.5 that $h(z) = z$ and so $\varphi_1 = \varphi_2$.

(14.6) Definition. Given three points z_1, z_2, z_3, not all the same, their ratio is define by

$$(z_1,z_2;z_3) = \frac{z_3 - z_1}{z_3 - z_2} .$$

If z_1, z_2, z_3, and z_4 are points, no three of which coincide, the cross ratio is defined by

$$(z_1,z_2;z_3,z_4) = \frac{(z_1,z_2;z_3)}{(z_1,z_2;z_4)}$$

$$= \frac{z_3 - z_1}{z_3 - z_2} \cdot \frac{z_4 - z_2}{z_4 - z_1}$$

The ratio of three points is invariant under translations, rotations, and dilations, as a simple calculation shows. The cross ratio is furthermore invariant under inversions and, hence, under all Möbius transformations.

(14.7) Definition: Given z_1, z_2 $\mathbb{D}$ we define the non-Euclidean distance between them by

$$d(z_1,z_2) = |\log(z_1,z_2;z_3,z_4)|$$

$$= \log \frac{1 + \left|\dfrac{z_1 - z_2}{1 - \bar{z}_1 z_2}\right|}{1 - \left|\dfrac{z_1 - z_2}{1 - \bar{z}_1 z_2}\right|} ,$$

where z_3 and z_4 are the points where $\partial\mathbb{D}$ intersects the circle through z_1 and z_2 that meets $\partial\mathbb{D}$ at right angles (z_3 is the one nearest z_2).

Since a holomorphic function preserves angles whenever its derivative is non-vanishing, the non-Euclidean distance is invariant under one-to-one conformal maps of $\mathbb{D}$ onto $\mathbb{D}$ because the cross ratio is.

The equality of the two expressions for $d(z_1,z_2)$ is easily proved when $z_1 = 0$ and $z_2 > 0$ (so $z_3 = 1$, $z_4 = -1$). Equality for the general case can then be shown by observing that both expressions are invariant under conformal one-to-one maps of $\mathbb{D}$ onto $\mathbb{D}$ and that the general case can be brought to the special case by such a map. It is left as an exercise to prove that d is a metric.

The disk with this metric is a geometry that satisfies all the axioms and postulates of Euclidean geometry except for the parallel postulate. (The "lines" in this geometry are arcs of circles which meet ∂D at right angles.)

(14.7) <u>Proof of the Riemann Mapping Theorem</u>. We will call a region G <u>logarithmically simply connected</u> if for each $f \in H(G)$ without zeros, there is a branch of $\log f$, that is, a function $g \in H(G)$ such that $f = \exp g$. We use the abbreviation log.s.c. We leave it as an exercise that simply connected (s.c.) domains are log.s.c. and conversely. Theorem 11.7 is the key element in the proof of this fact. The advantage of log.s.c. is that it is trivial to prove that if G is log.s.c. and $G' \cong G$, then G' is log.s.c. The corresponding result for s.c. is difficult to prove, using our difinition of s.c.

In short, our hypotheses are that G is log.s.c. and that $G \neq \mathbb{C}$.

(14.8) <u>Lemma</u>. G <u>is conformally equivalent to a subset of</u> $\mathbb{D}$.

<u>Proof</u>: For simplicity and without loss of generality, suppose $0 \in \mathbb{C}\setminus G$, so that there is a branch g of $\log z$ in $H(G)$: $\exp g(z) = z$, $z \in G$. Clearly g is one-to-one, for if $g(z) = g(z')$, then $z = \exp g(z) = \exp g(z') = z'$. Next, writing $G' = g(G)$, we show that $\mathbb{C}\setminus G'$ contains an open set. Now for any $z \in G'$, we must have $z + 2\pi i \notin G'$. Because if $g(z_1) = z$ and $g(z_2) = z + 2\pi i$ we get $z_1 = \exp z = \exp(z + 2\pi i) = z_2$, contradicting the single-valued nature of the function g. Thus G' and $G' + 2\pi i = \{z + 2\pi i : z \in G'\}$ are disjoint and $G' + 2\pi i$ is clearly open. By a translation and a dilation of $\mathbb{C}$ we may arrange that $\mathbb{D} \cap G" = \emptyset$ where $G" \cong G'$. We invert $G"$ to complete the proof--that is let $G^* = \{1/z : z \in G"\}$ and $G^* \subseteq \mathbb{D}$ with $G^* \cong G$. QED

(14.9) <u>Lemma</u>. <u>If</u> $G \subseteq \mathbb{D}$, $G \neq \mathbb{D}$ <u>and</u> $0 \in G$, <u>then there is a one-to-one conformal map</u> F <u>of</u> G <u>into</u> $\mathbb{D}$ <u>such that</u> $F(0) = 0$, $|F'(0)| > 1$, <u>and</u> $|F(z)| > |z|$ <u>for</u> $z \in G$, $z \neq 0$.

<u>Proof</u>: Choose $a \in \mathbb{D}\setminus G$, and let F_1 be a Blaschke factor that maps a to 0. Let F_2 be a branch of $\sqrt{z}$ that is holomorphic in $F_1(G)$. (Take $F_2(z) = \exp(\frac{1}{2}\log z)$, which exists since $F_1(G)$ is log.s.c.) Let F_3 be a Blaschke factor which takes $F_2(F_1(0))$ to 0. Define $F = F_3 \circ F_2 \circ F_1$. It is clear that F is one-to-one and holomorphic and takes G into D.

Define $\Phi_j = F_j^{-1}$, $j = 1,2,3$, and observe that $\Phi = \Phi_1 \circ \Phi_2 \circ \Phi_3$ satisfies the hypotheses of Schwarz's Lemma, so that $|\Phi'(0)| < 1$ and $|\Phi(z)| < |z|$ since $\Phi(z)$ cannot be $e^{i\lambda}z$. (It is not one-to-one in $\Phi(\mathbb{D}) = \mathbb{D}$.). The result now follows for F since $|F'(0)| = 1/|\Phi'(0)|$ and $|z| = |\Phi(F(z))| < |F(z)|$. QED

(14.10) We let $\mathcal{F}$ consist of all one-to-one conformal maps f of the region G into $\mathbb{D}$, such that $f(0) = 0$ and $f'(0) > 0$ (supposing for convenience that $0 \in G$); we also suppose that the function 0 belongs to $\mathcal{F}$. According to Hurwitz's Theorem (or rather its consequence Theorem 11.15) any converging sequence of univalent functions has a limit which is either univalent or constant. Thus $\mathcal{F}$ must be a closed set (the only constant function which can occur as a limit of functions in $\mathcal{F}$ is 0 because of the restriction $f(0) = 0$ on functions in $\mathcal{F}$). Since $\mathcal{F}$ is obviously bounded we conclude $\mathcal{F}$ is compact. Let $\varphi : \mathcal{F} \to \mathbb{R}$ be defined by $\varphi(f) = f'(0)$. Now φ is continuous on $\mathcal{F}$, so there is an $f_0 \in \mathcal{F}$ such that $\varphi(f_0) = \max\{\varphi(f) : f \in \mathcal{F}\}$. By the first lemma (14.8) we see that $\varphi(f_0) > 0$ since $\mathcal{F}$ contains at least one non-zero function. We claim that f_0 maps G onto $\mathbb{D}$. Otherwise, by the preceding lemma, there is a function F that maps $f_0(G)$ one-to-one conformally into $\mathbb{D}$, with $F'(0) > 1$. Then $F \circ f_0 \in \mathcal{F}$ and $\varphi(F \circ f_0) = F'(0)f_0'(0) > f_0'(0) = \varphi(f_0)$. This contradicts the choice of f_0. Thus f_0 must be onto and is the required mapping function. QED

An alternative proof is to maximize $|f(z_0)|$ over $f \in \mathcal{F}$ for some $z_0 \in G$, $z_0 \neq 0$. The other half of 14.9 is then used.

NOTES: A constructive proof of the Riemann Mapping Theorem occurs in [Carathéodory, vol. II, Sect. 3.15-3.23].

Exercises

1. If $\varphi : \hat{\mathbb{C}} \to \hat{\mathbb{C}}$ is one-to-one, onto and holomorphic (with the usual convention for holomorphic at ∞), prove that φ is a Moebius transformation.

2. Prove that d defined in (14.7) is a metric. Carry out the details of proving

$$|\log(z_1,z_2;z_3,z_4)| = \log \frac{1 + \left|\frac{z_1 - z_2}{1 - \bar{z}_1 z_2}\right|}{1 - \left|\frac{z_1 - z_2}{1 - \bar{z}_1 z_2}\right|}.$$

3. Show that if $f \in H(G)$, G simply connected and $Z(f) = \emptyset$. then f'/f has a primitive F with $e^F = f$. Conversely, if $e^F = f$ then F is a primitive of f'/f, so if G is not simply connected then $z - z_0$ does not have a branch of $\log(z - z_0)$ for some $z_0 \notin G$. (See Exercise 15 of Chapter 2 and Theorem 11.7).

4. Show that $\mathbb{C}^\wedge$ cannot be conformally equivalent to either $\mathbb{C}$ or $\mathbb{D}$ and that $\mathbb{C}$ cannot be conformally equivalent to $\mathbb{D}$.

5. Show that the Möbius transformation which takes -1 to 0, 0 to 1, and 1 to ∞, takes the unit circle (except 1) to the imaginary axis. Do this without computing the transformation; using only the fact that Möbius transformations take circles to circles or lines and preserve angles.

6. Find a Möbius transformation φ that maps $G = \{z : \text{Re } z > 0\}$ conformally onto $\mathbb{D}$ such that $\varphi(1) = 0$ and $\varphi'(1) > 0$. What if the first condition is changed to $\varphi(a) = 0$ for some other $a \in G$?

7. Find a conformal map from $G = \{z : |z| < 1 \text{ and } |z - (1 + i)| < 1\}$ onto $\mathbb{D}$. G is an American football-shaped figure with its ends at 1 and i where its bounding arcs make angles of $\frac{\pi}{2}$. (Hint: Use a Möbius transformation--keeping in mind that it will preserve angles except at its singularity--to map G onto the first quadrant $\{z : \text{Re } z > 0, \text{Im } z > 0\}$. Compose with maps from the first quadrant to a half-plane and the half-plane to the disk.)

8. Taking for granted that $-\log f(z)$ is <u>harmonic</u> where $f(z)$ is analytic and non-zero, show that for any simply connected G and $a \in G$, there is a real-valued harmonic function $U(z)$ such that $\lim_{z \to \partial G} U(z) = 0$ and $\lim_{z \to a} U(z) = +\infty$. (A function of two variables $u(x,y)$ is said to be <u>harmonic</u> if $\frac{\partial^2 U}{\partial x^2} + \frac{\partial^2 U}{\partial y^2}$ 0, but you do not need this to do the problem.

The function U--properly normalized--is known as the Green's function of G will pole at a.)

9. Use Schwartz's Lemma (14.3) to show that any conformal map φ of a domain G onto $\mathbb{D}$ such that $\varphi(a) = 0$ satisfies

$$|\varphi'(a)| < \frac{1}{\operatorname{dist}(a, \partial G)} .$$

If $\psi : \mathbb{D} \to G$ is the inverse of φ, show that $|\psi'(z)|(1 - |z|^2) < \operatorname{dist}(\psi(z), G)$.

10. Given a simply connected region G, what is the cardinality of all conformal one-to-one maps of G onto G?

11. What can be said about an entire function f such that $|f(z)| = 1$ whenever $|z| = 1$?

12. What can be said about an entire function f such that f(z) is real whenever $|z| = 1$?

*13. Is there a simply connected region G and a transcendental entire function f (i.e. not a polynomial) such that $f|_G$ maps G conformally one-to-one onto G?

14. If f and g map $\mathbb{D}$ conformally one-to-one onto $\mathbb{D}$, which of the following does the same?

a) $\frac{1}{2}(f + g)$,   b) fg,

c) $\sqrt{fg}$ (assuming f and g have a common zero),

d) $f*g$ where $\sum a_n z^n * \sum b_n z^n \equiv \sum a_n b_n z^n$.

15. Find the general form of all one-to-one conformal maps of S onto itself if S is the horizontal strip: $S = \{z : |\operatorname{Im} z| < 1\}$. Try other simply connected domains too.

16. Given domains G, G', G" and holomorphic maps $f : G \to G'$ and $g : G' \to G''$, let $H = g \circ f$. Suppose H is a one-to-one conformal map of G onto G". What can be said of g and f?

§15. Carathéodory Kernels and Farrell's Theorem

(15.1) Definition. Given a sequence $\{G_n\}$ of regions and a region $G \neq \emptyset$ such that $G \subseteq G_{n+1} \subseteq G_n$ for $n = 1,2,3,..$, we say that a superset G' of G is suitable if G' is connected and $G' \subseteq \cap G_n$. Then $\ker[G_n : G]$, the kernel of $\{G_n\}$ with respect to G, is defined as the union of all suitable supersets of G.

Note that $\ker[G_n : G]$ is that connected component of the interior of $\cap G_n$ which contains G.

(15.2) Lemma. If G' is a suitable superset of G then $\ker[G_n : G'] = \ker[G_n : G]$.

Proof: Any suitable superset of G' is a suitable superset of G. On the other hand, if G'' is a suitable superset of G, then $G' \cup G''$ is a suitable superset of G'. QED

(15.3) Lemma. If G' is another region contained in $\cap G_n$, and if $G \cap G' \neq \emptyset$, then $\ker[G_n : G] = \ker[G_n : G']$.

Proof: $G \cup G'$ is a suitable superset of both G and G' so $\ker[G_n : G] = \ker[G_n : G \cup G'] = \ker[G_n : G']$ by Lemma 15.2. QED

(15.4) Theorem (Kernel Convergence). Suppose G and $\{G_n\}$ are as in 15.1, that $0 \in G$, and that for some bounded region H, there are functions $\varphi_n \in H(G_n)$, with $\varphi_n : G_n \to H$ one-to-one and onto and $\varphi_n(0) = 0$. Write $\psi_n = \varphi_n^{-1}$ and suppose that $\psi_n \to \psi_0$ in $H(H)$. Then ψ_0 maps H one-to-one and onto $\ker[G_n : G]$.

Proof: From Hurwitz's Theorem (11.15) it follows that ψ_0 is either one-to-one or a constant. We first show that ψ_0 is not constant by showing $\psi_0'(0) \neq 0$. Now $\psi_n(H) \supseteq G$ and G contains a disk $D_\rho = \{z \in \mathbb{C} : |z| < \rho\}$. Also H is contained in some disk $D_R = \{z : |z| < R\}$. Applying Schwarz's Lemma (14.3) to $\frac{1}{R}\varphi_n(\rho z)$ on $\mathbb{D}$ shows that $|\varphi_n'(0)| < R/\rho$. Thus $|\psi_n'(0)| > \rho/R$ and hence $|\psi'(0)| > \rho/R$. Now, letting $G' = \psi_0(H)$ we show that G' is contained in $\cap G_n$ and $G' \cap G \neq \emptyset$. The latter is clear since $0 \in G' \cap G$. It is clear that $G' \subseteq \cap \bar{G}_n$. Let $z_0 \in G'$ and let C be a circle centered at $\psi_0^{-1}(z_0)$ which is contained, with its interior, in H. By Rouché's Theorem, for large enough n the function $\psi_n - z_0$ has the same number of zeros within C as does ψ_0, namely one. Thus $z_0 \in \psi_n(H) = G_n$.

Thus G' satisfies the hypotheses of Lemma 15.3 so $\ker[G_n : G'] = \ker[G_n : G]$. In particular $G' \subseteq \ker[G_n : G]$. To finish the proof we introduce some preliminary material.

(15.5) Definition. In an open set G, we say that the sequence $\{f_n\}$ of functions converges continuously in G to the function f_0, provided that for each sequence $\{z_n\}$ of points in G that converges to a point in G,

$$\lim_{n\to\infty} f_n(z_n) = f_0(\lim_{n\to\infty} z_n).$$

(15.6) Proposition. If $f_n \to f_0$ in $H(G)$, then f_n converges continuously to f_0.

Proof: Observe that, for any sequence $\{z_n\}$ in G with a limit z_0 in G, the set $K = \{z_n : n = 1,2,3,\ldots\} \cup \{z_0\}$ is compact. Thus $f_n \to f_0$ uniformly on K and hence

$$|f_n(z_n) - f_0(z_0)| \leq |f_n(z_n) - f_0(z_n)| + |f_0(z_n) - f_0(z_0)|$$

$$\leq \sup_K |f_n - f_0| + |f_0(z_n) - f_0(z_0)|.$$

Since f_0 is analytic, hence continuous, in G the second term tends to zero. The first tends to zero by definition of convergence in $H(G)$. QED

(15.7) Returning to the proof, write $K = \ker[G_n : G]$ and take any $w_0 \in K$. We need to show that $w_0 \in G' = \psi(H)$. Since $K \subseteq \cap G_n$, φ_n is defined on K. Let $\Phi_n = \varphi_n|_K$. Since H is bounded the family $\{\Phi_n\}$ is bounded in $H(K)$ and so we may select a subsequence Φ_{n_k} which converges in $H(K)$ to some Φ. Let $z_0 = \Phi(w_0)$, $z_k = \Phi_{n_k}(w_0)$ so that $z_k \to z_0$. Since $\psi_{n_k} \to \psi_0$ in $H(H)$ we have, by 15.6,

$$\psi_{n_k}(z_k) \to \psi_0(z_0)$$

Because $\psi_{n_k}(z_k) = \psi_{n_k}(\Phi_{n_k}(w_0)) = w_0$, we get $w_0 = \psi_0(z_0) \in \psi(H)$. QED

(15.8) Theorem. Suppose that the G_n are bounded and simply connected, that $G \subseteq G_{n+1} \subseteq G_n$ for $n = 1,2,3,\ldots$, and that $0 \in G$. Let $\varphi_n : G_n \to \mathbb{D}$ be the normalized mapping functions of G_n one-to-one onto the unit disk and let $\psi_n = \varphi_n^{-1}$. Then ψ_n converges in $H(\mathbb{D})$ to ψ_0, the inverse of the normalized mapping function φ_0 from $K = \ker[G_n : G]$ one-to-one onto $\mathbb{D}$. Moreover, if $\Phi_n = \varphi_n|_K$, then $\psi_0 \circ \Phi_n$ converges in $H(K)$ to the identity function.

Proof: By the preceding theorem, any limit point ψ in $H(\mathbb{D})$ will be one-to-one and onto K and satisfy $\psi(0) = 0$ and $\psi'(0) > 0$. Since the normalized mapping function is unique, $\psi = \psi_0$. Thus $\{\psi_n\}$ has at most one limit point. Since it is bounded it has at least one. Thus $\psi_n \to \psi_0$. Moreover, for $z \in K$, $\varphi_n(z) \to \varphi_0(z)$. Thus $\psi_0 \circ \Phi_n(z)$ converges to $\psi_0(\varphi_0(z)) = z$ for $z \in K$. QED

We now show how these ideas on conformal mapping can be used to prove theorems on approximation by proving a theorem of Farrell that exemplifies the technique.

(15.9) Definition. A sequence $\{f_n\}$ of functions in $H(G)$ is said to converge boundedly to a function f if there is a constant $M > 0$ such that $|f_n(z)| \leq M$ for all $z \in G$ and all n and $\lim_{n\to\infty} f_n(z) = f(z)$ for all $z \in G$.

It follows from the compactness principle that $f \in H(G)$ if $f_n \to f$ boundedly. For, some subsequence of $\{f_n\}$ converges in $H(G)$ and its limit must be f.

(15.10) Definition. If G is a bounded region let G^* be the complement of the closure of the unbounded component of the complement of the closure of G. Let $G^\#$ be the component of G^* that intersects G. Then $G^\#$ is called the inside of the outer boundary of G.

The outer boundary of G is the boundary of the unbounded component of $\mathbb{C}\setminus\bar{G}$. In the case of an annulus, this is the outer circle. Then G^* consists of those points "surrounded" by this outer boundary. For an annulus this is the disk which consists of the annulus and its hole. Finally $G^\#$ is the component of G^* containing G. To see that G^* may have more than one component, let G be a spiral ribbon that winds infinitely often around a disk, approaching its boundary. Then G^* consists of the ribbon and the disk, while $G^\#$ is just the ribbon.

(15.11) Farrell's Theorem. In order that f be the bounded limit of a sequence of polynomials in the bounded region G, it is necessary and sufficient that f have an extension F that is bounded and holomorphic in $G^{\#}$.

Proof: First, suppose that $\{p_n\}$ is a sequence of polynomials converging boundedly to f in G. The p_n are uniformly bounded in G, by hypothesis; hence they are uniformly bounded on ∂G. Now, the boundary of $G^{\#}$ is contained in ∂G, and so the p_n are uniformly bounded in $G^{\#}$. The compactness principle asserts that there is a subsequence $\{p_{n_k}\}$ that converges in $H(G^{\#})$ to a function F. Then F is bounded because the p_{n_k} are uniformly bounded, and certainly for $z \in G$, $F(z) = \lim p_{n_k}(z) = \lim p_n(z) = f(z)$.

In the other directions, we need some preliminary topological results.

(15.12) Definition. If G is bounded and $G = G^{\#}$ then G is called a Carathéodory region.

(15.13) Proposition. $(G^{\#})^{\#} = G^{\#}$, and therefore $G^{\#}$ is always a Carathéodory region.

(15.14) Proposition. $G^{\#}$ is simply connected.

The proofs are trivial applications of the definitions.

(15.15) Theorem. If G is a Carathéodory region, then there exists a sequence $\{G_n\}$ of simply connected regions such that $\bar{G} \subseteq G_{n+1} \subseteq G_n$, and such that $G = \ker[G_n : G]$.

Proof: Let $J_n = \{z : \text{dist}(z,G) < \frac{1}{n}\}$. The J_n need not be simply connected, as the reader can see by simple examples. Let $G_n = J_n^{\#}$. Let $K = \ker[G_n : G]$. Clearly, $G \subseteq K$. In the other direction, suppose there is a point in K that is not in G. Then there is a point $w \in K$, $w \notin \bar{G}$. If w is in the unbounded component of $\mathbb{C}\setminus\bar{G}$, let γ be a path connecting w to ∞ which doesn't meet $\bar{G}$. If $\frac{1}{n} < \text{dist}(\gamma,\bar{G})$ then γ is in the unbounded component of J_n, i.e. $\gamma \cap J_n^{*} = \emptyset$. Thus $\gamma \cap G_n = \emptyset$, contradicting $w \in K \subseteq G_n$ for all n. Thus $K \subseteq G^{*}$. But K is connected and intersects G, so it is contained in the component of G^{*} that contains G, i.e. $K \subseteq G^{\#}$. Since $G = G^{\#}$ we are done. QED

(15.16) To complete the proof, let F be the extension of f to $G^{\#}$. It suffices to show F is the limit of a uniformly bounded sequence of polynomials in $G^{\#}$. So, without loss of generality, we may suppose G is a Carathéodory region and let $\{G_n\}$ be domains whose existence was just shown. Let φ_n be the normalized mapping function of G_n onto $\mathbb{D}$ and let φ_0 be the normalized mapping function of G onto $\mathbb{D}$. Let $\psi_0 = \varphi_0^{-1}$ and let $F_n = f \circ \psi_0 \circ \varphi_n$. We know (15.8) that $\psi_0 \circ \varphi_n(z) \to z$ for $z \in G$, so $F_n \to f$ boundedly on G. The F_n are holomorphic in G_n and $\bar{G}$ is a compact subset of G_n, so Runge's Theorem asserts the existence of polynomials p_n such that $\|p_n - F_n\|_{\bar{G}} < \frac{1}{n}$. Clearly, $\{p_n\}$ is uniformly bounded on G and conberges on G to the same limit as $\{F_n\}$, namely f. QED

NOTES: Farrell's Theorem (15.11) occurs in [O. J. Farrell]. An extension of Farrell's Theorem to disconnected G can be found in [L. A. Rubel and A. L. Shields]. The Kernel Convergence Theorem 15.4 is a special case of a theorem by Carathéodory on the convergence of regions and conformal mapping functions. See [C. Carathéodory, 2, Chapter 5, pp. 120-123].

Exercises

1. Compute $\ker[G_n : G]$ in the following situations.

   a) G is the square $-2 < \operatorname{Re} z < 0$, $|\operatorname{Im} z| < 1$. G_n are the rectangles $-2 - \frac{1}{n} < \operatorname{Re} z < 2$, $|\operatorname{Im} z| < 1 + \frac{1}{n}$.

   b) Let F be the "figure eight": $F = \{z : |z - 1| \le 1\} \cup \{z : |z + 1| \le 1\}$. Let $G = \{z : |z| < 1\}$ and let $G_n = \{z : \operatorname{dist}(z,F) < \frac{1}{n}\}$.

   c) Let R be the rectangle $-2 < \operatorname{Re} z < 2$, $|\operatorname{Im} z| < 1$. Let Γ_k be the line segment from $-\frac{1}{k} + (1 - \frac{1}{k})i$ to $-\frac{1}{k} - (1 - \frac{1}{k})i$. Define $G = \{z : |z - 1| < 1\}$ and $G_n = R \setminus \bigcup_{k=1}^{n} \Gamma_k$.

2. Construct an example to show that $\ker[G_n : G]$ may be simply connected even when neither G nor any of the G_n is simply connected.

3. Show that if a sequence of polynomials $\{p_n\}$ converges boundedly to a function $f \in H(G)$, then $\{p_n\}$ converges to f in $H(G)$, that is, uniformly on compact subsets of G.

4. Prove directly (without kernel convergence) that if f is a bounded analytic function on $\mathbb{D}$ then f is a bounded limit of polynomials. There are two well-known methods. a) Consider $f_r(z) = f(rz)$, $r < 1$, and b) consider $p_n(z) = \sigma_n(f,z) = \sum_{k=0}^{n} (1 - \frac{k}{n}) a_k z^k$ when $f(z) = \sum_k a_k z^k$. The latter will be familiar to those who have studied Fourier series and their Cesaro means.

5. Prove Propositions 15.13 and 15.14.

§16. Ring (not Algebra) Isomorphisms of $H(G)$

We return here to the ring structure of $H(G)$. A ring homomorphism of R_1 to R_2 is a function $\varphi : R_1 \to R_2$ which preserves multiplication i.e. $\varphi(rs) = \varphi(r)\varphi(s)$ and $\varphi(r + s) = \varphi(r) + \varphi(s)$ for all r, $s \in R_1$. A ring isomorphism is a ring homomorphism that is one-to-one and onto. If G and G' are two conformally equivalent domains in $\mathbb{C}$ then there is an algebra isomorphism from $H(G)$ to $H(G')$ as we saw in Chapter 5. An algebra isomorphism will be a ring isomorphism which additionally preserves scalar multiplication. It follows from Proposition 5.4 that $H(G)$ and $H(G')$ are isomorphic as algebras if and only if G and G' are conformally equivalent. But a ring isomorphism can exist without conformal equivalence.

(16.1) Definition. A function φ on G is anticonformal if $\bar{\varphi}$ is conformal. Two regions G and G' are anticonformally equivalent if there is an anticonformal map φ from G onto G'.

If $\varphi : G \to G'$ is anticonformal and onto then there is a ring isomorphism $f \longmapsto f^*$ from $H(G')$ to $H(G)$ defined by $f^*(z) = \overline{f(\varphi(z))}$. It follows from the power series representation of f that $f^* \in H(G)$: If $f(w) = \sum_{n=0}^{\infty} a_n(w - w_0)^n$ is the expansion of f about w_0, then $f^*(z) = \sum_{n=0}^{\infty} \bar{a}_n(\overline{\varphi(z)} - \bar{w}_0)^n$ which is holomorphic because $\bar{\varphi}$ is. It is clear that $f \longmapsto f^*$ is a ring isomorphism but not an algebra isomorphism because $(af)^* = \bar{a}f^*$, for $a \in \mathbb{C}$.

Our main theorem on ring isomorphisms is due to Lipman Bers.

(16.2) Theorem. (Bers) Let G and Γ be two regions in $\mathbb{C}$. Then $H(G)$ and $H(\Gamma)$ are ring-isomorphic if and only if G and Γ are either conformally or anticonformally equivalent. Moreover, if $f \longmapsto f^*$ is a ring isomorphism then it is induced by a conformal or anticonformal map.

This means that either there exists a conformal $\varphi : G \to \Gamma$ such that $f = f^* \circ \varphi$ or else there exists an anticonformal $\varphi : \Gamma \to G$ such that $f = (f^* \circ \varphi)^-$.

Proof: Consider a fixed isomorphism $f \longmapsto f^*$ from $H(G)$ to $H(\Gamma)$. It takes a function $f(z)$, $z \in G$ into a function $f^*(\zeta)$, $\zeta \in \Gamma$ and a set $S \subseteq H(G)$ into a set $S^* \subseteq H(\Gamma)$. Let c be a complex number. We also use the letter c to denote the constant function with value c, whatever its domain. We say c is rational if $\operatorname{Re} c$ and $\operatorname{Im} c$ are both rational.

(16.3) Lemma. Either $i^* = i$ and then for every rational constant r, $r^* = r$, or $i^* = -i$ and $r^* = \bar{r}$.

Proof: $i^* \cdot i^* = (-1)^* = -1$ so $i^* = \pm i$. Suppose $i^* = i$; the alternative can be treated similarly. Then $r^* = r$ if r has integer coordinates. If r is rational then nr has integer parts for some n. Then $nr^* = (nr)^* = nr$ gives $r^* = r$. QED

(16.4) Lemma. If c is a constant, then so is c^*.

Proof: This is clear if c is rational because of (16.3). If c is not rational, then $c - r$ is invertible in $H(G)$ for every rational r. Consequently $c^* - r^*$ is invertible in $H(\Gamma)$. Thus c^* can take no rational values and must be constant because of the open mapping principle. QED

From now on we shall assume $i^* = i$; the alternative (namely $i^* = -i$) can be treated the same way. Let I_a denote the ideal of functions in $H(G)$ that vanish at a point $a \in G$, with a similar notation $I_\alpha \subseteq H(\Gamma)$ when $\alpha \in \Gamma$.

(16.5) Lemma. There exists a one-to-one map $z \mapsto \zeta = \varphi(z)$ of G onto Γ such that $(I_a)^* = I_{\varphi(a)}$.

Proof: We remark that much of the work of Bers' Theorem will be showing that φ is holomorphic--something we don't know at this stage. If $f \in H(G)$ and (f) denotes the principal ideal generated by f, then $(f)^* = (f^*)$ and (f^*) is a maximal ideal if and only if (f) is. Now $I_a = (f)$ with $f(z) = z - a$, and is maximal. Thus $(I_a)^* = (f_a^*)$, so $(I_a)^*$ is a maximal principal ideal. By combining (13.7) and (13.9) we see that $(I_a)^* = I_\alpha$ for some α. Define $\varphi(a) = \alpha$. Because $*$ is an isomorphism φ will be one-to-one and onto. QED

(16.6) Lemma. For every $z_0 \in G$ and $f \in H(G)$, $f(z_0)^* = f^*(\varphi(z_0))$. That is, the constant function $f(z_0)$ corresponds to $f^*(\varphi(z_0))$ under $*$.

<u>Proof</u>: Let $c = f(z_0)$. Then $c - f \in I_{z_0}$ so that $c^* - f^* \in I_{\varphi(z_0)}$. This means that $c^* - f^*(\varphi(z_0)) = 0$. QED

The following two lemmas follow immediately from (16.6) and the preceeding lemmas.

(16.7) <u>Lemma</u>. <u>If</u> $z_0 \in G$ <u>and</u> $f(z_0)$ <u>is a rational number, then</u> $f(z_0) = f^*(\varphi(z_0))$.

(16.8) <u>Lemma</u>. <u>If</u> $f \in H(G)$ <u>is univalent then</u> f^* <u>is univalent</u>.

<u>Proof</u>: If $\alpha \in \Gamma$, let $a \in G$ with $\varphi(a) = \alpha$. Now $(f - f(a)) = I_a$ so $(f^* - f(a)^*) = (f^* - f^*(\alpha)) = I_\alpha$. Thus $f^*(\zeta) = f^*(\alpha)$ only if $\zeta = \alpha$. QED

(16.9) <u>Lemma</u>. <u>Let</u> $f(z)$ <u>be univalent in</u> G, <u>let</u> $\{w_n\}$ <u>be a sequence of distinct rational points in</u> $f(G)$ <u>converging to a point</u> $w \neq \infty$, <u>and let</u> $\{z_n\}$ <u>be the sequence in</u> G <u>with</u> $w_n = f(z_n)$. <u>Then</u> w <u>is a boundary point of</u> $f(G)$ <u>if and only if there exists a function</u> g <u>in</u> $H(G)$ <u>such that</u> $g(z_n) = n$ <u>for</u> $n = 1,2,3,\ldots$.

<u>Proof</u>: If w is an interior point of $f(G)$ then $z_n = f^{-1}(w_n)$ converges to $f^{-1}(w)$. Consequently $\{g(z_n)\}$ has a finite limit for all $g \in H(G)$. Conversely if w is a boundary point of $f(G)$, then $\{z_n\}$ is admissible in G (Definition (12.1)). For if $\{z_n\}$ had a limit point z, then $f(z) = w$ would contradict the location of w on the boundary. Consequently, the Interpolation Theorem (12.14) can be applied to produce g. QED

(16.10) <u>Lemma</u>. <u>Let</u> $f(z)$ <u>be a univalent function in</u> G <u>so that</u> $f^*(\zeta)$ <u>is univalent in</u> Γ. <u>Then the sets</u> $f(G)$ <u>and</u> $f^*(\Gamma)$ <u>are identical</u>.

<u>Proof</u>: It follows from (16.7) that $f(G)$ and $f^*(\Gamma)$ contain exactly the same rational points, so at least their closures are the same. Let $w \in \partial f(G)$ and let w_n be a sequence of rational points in $f(G)$ converging to w. Let $h = f^{-1}$, $k = (f^*)^{-1}$, $z_n = h(w_n)$ and and $\zeta_n = k(w_n)$. Now $f(z_n) = f^*(\varphi(z_n)) = w_n$ so $\zeta_n = \varphi(z_n)$. From Lemma 16.9 there exists $g \in H(G)$ with $g(z_n) = n$, $n = 1,2,3,\ldots$. Then $g^*(\zeta_n) = g^*(\varphi(z_n)) = g(z_n)^* = n$. Thus, again by Lemma 16.9, w is a boundary point of $f^*(\Gamma)$. Reversing the roles of f and f^*

shows that every boundary point of $f^*(\Gamma)$ is a boundary point of $f(G)$. Thus $f(G)$ and $f^*(\Gamma)$ are open sets with the same closure and the same boundary. Thus, they are identical. QED

Note that Lemma 16.10 implies that G and Γ are conformally equivalent. To delimit the nature of the isomorphism $*$ further, we need the following result, whose proof is given in several steps.

(16.11) Lemma. For every constant c, $c^* = c$.

For any set $B \subseteq \mathbb{C}$ we denote by $m[B]$ the set of all complex numbers d such that $z + d \in B$ whenever $z \in B$. That is, the translation by d maps B into B.

(16.12) Lemma. For every univalent function $f \in H(G)$ and for every constant c, the difference $c - c^*$ belongs to $m[f(G)]$.

Proof: The function $f_1 = f + c$ is univalent in G and the functions f^* and $f_1^* = f^* + c^*$ are univalent in Γ, by Lemma 16.9. By virtue of Lemma 16.10 $f(G) = f^*(\Gamma)$ and $f_1(G) = f_1^*(\Gamma)$. But a translation by c takes $f(G)$ into $f_1(G)$ and a translation by $-c^*$ takes $f_1^*(\Gamma) = f_1(G)$ into $f^*(\Gamma) = f(G)$. Thus, a translation by $c - c^*$ take $f(G)$ into itself. QED

Proof of Lemma 16.11. Assume, on the contrary, that $c - c^* = re^{i\theta}$, $r > 0$. If $G \neq \mathbb{C}$ then there exists either a finite line segment in G with its endpoints on ∂G or a semi-infinite line with its endpoint in ∂G. In either case there exists a linear fractional transformation $f \in H(G)$ so that $f(G)$ contains $\{te^{i\theta} : t < 0\}$ and $0 \notin f(G)$. But this contradicts the Lemma 16.12. This leaves only the case $G = \mathbb{C}$, which is handled by the next lemma. QED

(16.13) Lemma. If $G = \mathbb{C}$ (so that $\Gamma = \mathbb{C}$), then $c_n \to \infty$ implies $c_n^* \to \infty$.

Proof: Set $f(z) = z$. Then $f^*(\zeta)$ is univalent in $\mathbb{C}$, and hence $f^*(\zeta) = A\zeta + B$ for some constants A and B, $A \neq 0$. (See Exercise 17, Chapter 2.) For this f and for $z_0 = c$, Lemma 16.6 yields $c^* = A\varphi(c) + B$. So we only need to show $\varphi(c_n) \to \infty$ whenever $c_n \to \infty$. But if $c_n \to \infty$, then there exists a function $g \in H(G)$ such that $g(c_n) = n$. By Lemma 16.7, $g^*(\varphi(c_n)) = n^* = n \to \infty$. It follows that

$\varphi(c_n)$ cannot have a limit point in Γ so $\varphi(c_n) \to \infty$. QED

It follows from this lemma that $c \longrightarrow c^*$ is continuous on $\mathbb{C}$. (For if $c_n \to c$ then $\frac{1}{c_n^* - c^*} \to \infty$.) But $*$ is the identity at rational points so $*$ is the identity everywhere. This proves (16.11). Putting Lemmaa 16.6 and 16.11 together gives

(16.14) Lemma. For every $z_0 \in G$, $f(z_0) = f^*(\varphi(z_0))$

This would prove the theorem if we knew φ was holomorphic in G. To show this select $f(z) = z$ in Lemma 16.14. This gives $z = f^*(\varphi(z))$, so that φ is inverse to the univalent function f^* which makes it holomorphic. (See Exercise 11, Chapter 2).

NOTES: The reference for Bers' Theorem (16.2) is [L. Bers]. See also [H. Iss'sa].

Exercises

1. Show that $H(\mathbb{D})$ and $H(\mathbb{A})$ are not ring isomorphic where $\mathbb{A}$ is an annulus $r_1 < |z| < r_2$. Is there a topological vector space isomorphism between $H(\mathbb{D})$ and $H(\mathbb{A})$?

2. Prove that two annuli are not conformally equivalent unless they are similar.

3. Let $\mathbb{A} = \{z : 1 < |z| < 2\}$ and

$$\mathbb{A}' = \{z : 1 < |z| < 3\}.$$

Find an algebraic property of $H(\mathbb{A})$ that $H(\mathbb{A}')$ does not have. Consider both ring properties and algebra properties.

*4. Is $H(\mathbb{D})$ isomorphic to either $H(\mathbb{D} \times \mathbb{D})$ or $H(\mathbb{B})$, where $\mathbb{B} = \{(z,w) \in \mathbb{C}^2 : |z|^2 + |w|^2 \leq 1\}$?

5. Does there exist a region G such that $H(G)$ is ring isomorphic to $H(G\setminus\{z_0\})$ for some point $z_0 \in G$? *Can such regions be characterized?

*6. Suppose $f \in H(\mathbb{D})$ and suppose its range is dense in $\mathbb{C}$. Does there exist a bounded $\beta \in H(\mathbb{D})$ such that the range of $f + \beta$ is $\mathbb{C}$?

*7. Suppose A and A' are arbitrary (not necessarily open) sets in $\mathbb{C}$ and suppose that the rings of germs $H(A)$ and $H(A')$ are isomorphic. Can any conclusions about A and A' be drawn?

*8. For a region G, let $S(G)$ be the topological semigroup of all holomorphic functions $f : G \to G$ with the operation of composition. Suppose $S(G)$ is isomorphic to $S(G')$. What then?

9. Find a sequence $G_1 \subset G_2 \subset \cdots$ of bounded regions so that $\cup G_i = \mathbb{C}$ and such that $H(G_i) \not\cong H(G_j)$, $i \neq j$.

10. Can a Carathéodory domain be conformally equivalent to a non-Carathéodory domain?

*11. Suppose G is a region such that if Λ is a straight line intersecting G, then Λ cuts G into two regions G' and G'' so that $H(G') \cong H(G'')$. What can you say about G? (For example, G cannot be an annulus. If G is simply connected then must G be convex?)

§17. Dual Space Topologies

This chapter is intended as a prerequisite for later chapters. In it we introduce a topology on the dual of a topological vector space. We present some of the standard results in the theory of Fréchet spaces and some additional results on topological vector spaces in general.

It is a common requirement that the topology on a topological vector space be Hausdorff. Henceforth we make this assumption.

(17.1) Definition. If E is a vector space and Γ is a vector space of linear functionals on E, then the weakest topology on E that makes every functional in Γ continuous is called the $\sigma(E,\Gamma)$ topology. (Or sometimes the Γ-topology.)

A neighborhood of 0 in the Γ-topology of E is any set N that contains a set of the form

$$N(\varepsilon,\gamma_1,\gamma_2,\ldots,\gamma_k) = \{x \in E : |\gamma_j(x)| < \varepsilon,\ j = 1,2,\ldots,k\}$$

for some finite collection $\gamma_j \in \Gamma$. Neighborhoods of other points $x \in E$ are obtained by translating neighborhoods of 0.

If we apply this definition to E^* in place of E with $\Gamma = E$ we get the weak*-topology. Any element $x \in E$ can be considered a linear functional on E^*: the one that takes $L \in E^*$ to $L(x) \in \mathbb{C}$.

(17.2) Definition. The weak*-topology on E^* is the $\sigma(E^*,E)$ topology. It is the weakest topology that makes the evaluations $L \longrightarrow L(x)$ continuous for all $x \in E$.

In particular, a sequence $\{L_n\}$ in E^* converges in the weak*-topology to $L \in E^*$ if and only if $L_n(x) \to L(x)$ for all $x \in E$. (In general E^* with $\sigma(E^*,E)$ is not metrizable so sequences and their convergence are not sufficient to define all topological concepts, such as closure of sets. One of the deeper results here will be the Banach-Dieudonné Theorem which says that, in certain circumstances, closure is equivalent to sequential closure.)

(17.3) Theorem. Let E be locally convex. Then the weak*-topology on E^* is a Hausdorff locally convex topology.

This follows, in fact, from a stronger result about $\sigma(E,\Gamma)$

topologies. We will say that Γ is <u>total</u> if $\gamma(x) = 0$ for all $\gamma \in \Gamma$ implies that $x = 0$. This implies that Γ separates points of E: If $x, y \in E$, $x \neq y$, then there is a $\gamma \in \Gamma$ such that $\gamma(x) \neq \gamma(y)$.

(17.4) <u>Theorem</u>. *If E is a vector space and Γ is a total space of linear functionals then $\sigma(E,\Gamma)$ is a Hausdorff locally convex topology and $E^* = \Gamma$. (More precisely, $(E, \sigma(E,\Gamma))^* = \Gamma$.)*

<u>Proof</u>: Let $x_0, y_0 \in E$, $x_0 \neq y_0$. Then there exists $\gamma \in \Gamma$ such that $\gamma(x_0) \neq \gamma(y_0)$. Let $r = \frac{1}{2}|\gamma(x_0) - \gamma(y_0)|$. Then $\{x \in E : |\gamma(x) - \gamma(x_0)| < r\}$ and $\{x : |\gamma(x) - \gamma(y_0)| < r\}$ are disjoint neighborhoods of x_0 and y_0 respectively. It is easy to verify that, for $\gamma_1, \gamma_2, \ldots, \gamma_k \in \Gamma$, the function $p(x) = \max\{|\gamma_1(x)|, |\gamma_2(x)|, \ldots, |\gamma_k(x)|\}$ is a seminorm and

$$N(\varepsilon, \gamma_1, \ldots, \gamma_k) = \{x : p(x) < \varepsilon\}$$

so these seminorms generate the topology of E.

To finish the proof we need the following lemma.

(17.5) <u>Lemma</u>. *If $\ker \gamma \supset \bigcap_{i=1}^{k} \ker \gamma_i$ then γ is a linear combination of $\gamma_1, \ldots, \gamma_k$.*

Proof: Define a mapping $T : E \to \mathbb{C}^k$ by $T(x) = (\gamma_1(x), \gamma_2(x), \ldots, \gamma_k(x))$. Define a linear functional L on $T(E)$ by $L(T(x)) = \gamma(x)$. This is well-defined because if $T(x) = T(y)$, then $x - y \in \bigcap_{i=1}^{k} \ker \gamma_i$ and so our hypothesis implies $\gamma(x) = \gamma(y)$. Use the Hahn-Banach Theorem to extend L to all of $\mathbb{C}^k$. It is elementary linear algebra that any linear functional on $\mathbb{C}^k$ has the form

$$L(z_1, z_2, \ldots, z_k) = \sum_{j=1}^{k} a_j z_j$$

for some scalars $a_j \in \mathbb{C}$. Thus we have $\gamma(x) = \sum a_j \gamma_j(x)$ as required.

To finish the proof of (17.4), suppose L is a continuous linear functional on E. Then $\{x \in E : |L(x)| < 1\}$ is a neighborhood of 0 so it must contain a set of the form $\{x \in E : |\gamma_j(x)| < \varepsilon, j = 1, 2, \ldots, k\}$ with $\gamma_j \in \Gamma$. Clearly then, if $y \in \bigcap_{j=1}^{k} \ker \gamma_j$ then $|L(y)| < 1$. Since this must also hold for ny for every n, we get $|L(y)| < n^{-1}$ and so $L(y) = 0$, i.e. $y \in \ker L$. By Lemma 17.5, L is a linear combination of $\gamma_1, \gamma_2, \ldots, \gamma_k$ and so belongs to Γ. QED

The following is a simple criterion for a linear functional on a topological vector space to be continuous.

(17.6) Proposition. Let L be a linear functional on a topological vector space E. Then L is continuous if and only if its kernel, ker L, is closed.

Proof: If L is continuous then ker L is $L^{-1}(\{0\})$. But the inverse image of a closed set is closed.

Now suppose ker L is closed and not equal to E. There exists an $x_0 \in E$ with $L(x_0) = 1$. The set $M = \{x \in E : L(x) = 1\}$ satisfies $M = x_0 + \ker L$ so M is closed. There is a neighborhood N of 0 such that $N \cap M = \emptyset$. We claim there is a neighborhood V of 0 such that $V \subseteq N$ and $\lambda x \in V$ whenever $x \in V$ and $|\lambda| \leq 1$. To see this, use continuity of scalar multiplication to obtain an open set $U \subset E$ containing 0 and an $\varepsilon > 0$ such that $\lambda x \in N$ whenever $x \in U$ and $|\lambda| \leq \varepsilon$. Let $V = \bigcup_{|\lambda| \leq \varepsilon} \lambda U$ which is a union of open sets and is therefore open. We claim there is an $\alpha > 0$ such that $V \subset L^{-1}(\{z : |z| < \alpha\})$. In fact, take $\alpha = 1$ and suppose there is an $x \in V$ with $|L(x)| \geq 1$. Multiply x by a scalar λ with $|\lambda| \leq 1$ so that $L(\lambda x) = 1$. But $\lambda x \in V \subset N$ and N does not meet M so $L(\lambda x) \neq 1$. This contradiction shows that $|L(x)| < 1$ for all $x \in V$. Consequently $|L(x)| < \varepsilon$ for all $x \in \varepsilon V$ and this means L is continuous at 0. Clearly this implies L is continuous everywhere. QED

(17.7) Corollary. Every one-dimensional topological vector space is isomorphic and homeomorphic to $\mathbb{C}$.

Proof: Let E be one-dimensional and let $\{x_0\}$ be a basis. The mapping $\varphi : \mathbb{C} \to E$ defined by $\varphi(\alpha) = \alpha x_0$ is clearly an isomorphism and continuous. Define $L = \varphi^{-1}$ so $L(\alpha x_0) = \alpha$. This is a linear functional with $\ker L = \{0\}$. Since E is Hausdorff, $\{0\}$ is closed and so L is continuous. QED

(17.8) Proposition. Every finite dimensional vector subspace of a topological vector space is closed and homeomorphic to $\mathbb{C}^n$ for some n.

Proof: We leave it to the reader to show first, by induction using

(17.6) and (17.7) that every n-dimensional topological vector space is isomorphic and homeomorphic to $\mathbb{C}^n$. Now suppose F is an n-dimensional subspace of a topological vector space E and let x be in the closure of E. For every neighborhood V of 0 in E there is an element $f_V \in F \cap (x + V)$.

Let $\Phi : F \to \mathbb{C}^n$ be an isomorphism and homeomorphism and let $\alpha_V = \Phi(f_V)$. For any neighborhood $F \cap V$ of 0 in F there exists a neighborhood $F \cap V'$ of 0 such that

$$F \cap (V' - V') \equiv \{x \in F : x = x_1 - x_2, x_i \in V'\} \subset F \cap V.$$

Consequently $f_U - f_W \in F \cap V$ whenever U, W are neighborhoods of zero with $U \subset V'$ and $W \subset V'$. It is now easy to verify that there is a unique $\alpha \in \mathbb{C}^n$ with the following property: For every $\varepsilon > 0$ there is a V such that $\|\alpha_U - \alpha\| < \varepsilon$ whenever $U \subset V$. in fact, using Φ to transfer the previous discussion to $\mathbb{C}^n$, for every neighborhood B of 0 in $\mathbb{C}^n$ there is a V' such that $\alpha_U - \alpha_W \in B$ whenever U, $W \subset V'$. If $B = B(0,\varepsilon)$ then $\alpha_V \in B(\alpha_U,\varepsilon)$ for all sufficiently small V and U. It is then easy to get a sequence of balls $B(\alpha_{U_n}, 1/n)$ which shrink down to an α. It now follows that $\Phi^{-1}(\alpha) = x \in F$. QED

Note to the sophisticated reader: We have produced a Cauchy net in F and then used completeness of F to get a limit for it.

We turn now to the consideration of a class of spaces that includes both $H(G)$ and $C(G)$ for G an open set in $\mathbb{C}$.

(17.9) Definition. A Fréchet space E is a complete metrizable locally convex topological vector space. For our purposes this means the following: The topology on E is given by a sequence $\{p_n : n = 1,2,\ldots\}$ of seminorms satisfying $p_n(x) \le p_{n+1}(x)$ for all $n \ge 1$ and all $x \in E$. Moreover if $\rho(x,y)$ is defined by

$$\rho(x,y) = \sum_{n=1}^{\infty} 2^{-n} \frac{p_n(x - y)}{1 + p_n(x - y)}$$

then ρ is a complete metric.

A sequence $\{\chi_k\}$ in a locally convex space E is said to be Cauchy if for every $\varepsilon > 0$ and every continuous seminorm p there is an integer k_0 such that $p(x_k - x_\ell) < \varepsilon$ whenever $k, \ell > k_0$. If the topology on E is given by a countable family of seminorms then E is called complete if every Cauchy sequence converges. It can be shown that E is complete if and only if the ρ defined in (17.9) is a complete metric. In fact a sequence is ρ-Cauchy if and only if it is p_n-Cauchy for each p_n and it is ρ-convergent if and only if it converges in the p_n sense.

The reader has no doubt observed the analogy with Chapter 3, especially (3.3) and (3.4).

(17.11) Theorem (Principle of Uniform Boundedness or Principle of Equicontinuity). *Let* I *be an index set and, for each* $i \in I$, *let* T_i *be a continuous linear operator from a Fréchet space* (X,ρ) *to a Fréchet space* (Y,η). *Suppose that for each* $x \in X$ *the set* $\{T_i(x) : i \in I\}$ *is bounded (see* 3.11). *Then* $\lim_{x\to 0} T_i(x) = 0$ *uniformly for* $i \in I$.

The meaning of this last statement is that for any $\varepsilon > 0$ there is a $\delta > 0$ such that

$$\eta(T_i(x),0) < \varepsilon \quad \text{whenever} \quad \rho(x,0) < \delta. \tag{17.12}$$

In particular, the δ does not depend on i. The reader should observe that this principle is special to linear operators and fails badly unless some kind of linearity or approximate linearity is present.

Proof: For given $\varepsilon > 0$ and each positive integer k let

$$X_k = \{x : \eta(\tfrac{1}{k} T_i(x),0) \le \varepsilon, \quad \text{all} \quad i \in I\}.$$

By continuity each X_k is closed and our assumption of boundedness implies $X = \bigcup_{k=1}^{\infty} X_k$. By the Baire Category Theorem (1.19), some X_{k_0} contains an open ball $B(x_0,\delta) = \{x \in X : \rho(x,x_0) < \delta\}$. This means if $\rho(x) < \delta$ then

$$\eta(\tfrac{1}{k_0} T_i(x_0 + x),0) \le \varepsilon$$

Thus

$$\begin{aligned}\eta(\tfrac{1}{k_0} T_i(x),0) &\le \eta(\tfrac{1}{k_0} T_i(x_0 + x),0) + \eta(\tfrac{1}{k_0} T_i(x_0),0) \\ &\le 2\varepsilon.\end{aligned}$$

Thus, $\sup_i \eta(T_i(x),0) \to 0$ as $\rho(x,0) \to 0$. QED

We now make use of some of the concepts presented so far in the case of $H(G)$. The ρ defined in (3.3) renders $H(G)$ a Fréchet space. We will use the Principle of Uniform Boundedness to characterize the weak* convergence of a sequence in $H(G)^*$.

(17.13) Theorem. If L_n is a sequence in $H(G)^*$ then $L_n(f) \to 0$ for every $f \in H(G)$ if and only if there is a compact set $K \subset G$ such that the corresponding germs $F_n = \Phi(L_n)$ in $H_0(\hat{\mathbb{C}} \setminus G)$ have analytic extensions to $\hat{\mathbb{C}} \setminus K$ and $F_n \to 0$ uniformly on compact subsets of $\hat{\mathbb{C}} \setminus K$.

We remark that this refers only to sequences in $H(G)^*$, not to nets or other types of convergence.

Proof: We apply (17.11) to the spaces $X = H(G)$ and $Y = \mathbb{C}$. The family of operators is $L_n : H(G) \to \mathbb{C}$ and the hypothesis $L_n(f) \to 0$ implies $\{|L_n(f)| : n \geq 1\}$ is bounded for each f. We can therefore conclude that

$$\sup_n |L_n(f)| \to 0 \quad \text{as} \quad f \to 0. \tag{17.14}$$

We claim this implies the existence of K.

Let $\{K_m\}$ be an exhaustion of G by compact sets and suppose for each m there is a function $f_n \in H(G)$ such that

$$\sup_n |L_n(f_m)| \geq m\|f_m\|_{K_m}$$

Then $g_m \equiv f_m/(m\|f_m\|_{K_m}) \to 0$ but $\sup_n |L_n(g_m)| \geq 1$.

This contradicts (17.14). Consequently, there is an m_0 such that

$$|L_n(f)| \leq m_0\|f\|_{K_{m_0}}, \quad \text{all } f \in H(G). \tag{17.15}$$

Take $K = K^G_{m_0}$ (Definition 11.2). The inequality (17.15) implies that each L_n has an extension to a continuous linear functional μ_n on $C(K)$. From the proof of the main duality theorem in Chapter 9, $F_n(w) = \int (z - w)^{-1} d\mu_n(z)$ which is analytic off K. Finally, for $w \notin K$, Runge's Theorem implies $(w - z)^{-1}$ can be approximated uniformly on K by elements of $H(G)$. This shows that $F_n(w) \to 0$ pointwise in

$\mathbb{C}^\wedge \setminus K$ (because $L_n(f) \to 0$, $f \in H(G)$). Inequality (17.15) shows that $\{F_n\}$ is bounded on compact subsets of $\mathbb{C}^\wedge \setminus K$ and so $F_n \to 0$ uniformly on compact subsets of $\mathbb{C}^\wedge \setminus K$.

The converse is an easy exercise.

Having obtained a criterion for convergence of sequences in $H(G)^*$ we turn to the task of reducing certain topological questions to convergence of sequences. Our goal is the following:

(17.16) Theorem. Let E be a separable Fréchet space. Then a linear subspace M of E^* is closed in the weak*-topology if and only if it is sequentially closed, i.e. iff whenever $\{L_n\}$ is a sequence in M converging weak* to $L \in E^*$, it follows that $L \in M$.

(Note that this definitely does not say that a functional in the weak*-closure of a subspace can be approached by a sequence in that subspace. It merely says that if those which can be so approached all belong to the subspace, then it is closed.)

(17.17) Definition. If $V \subset E$ then the polar of V, denoted $V°$, is the subset of E^* defined by $V° = \{L \in E^* : |L(x)| \le 1$ for all $x \in V\}$.

(17.18) Lemma. If E is separable and if V is a neighborhood of 0 in E, then $V°$ with the relative weak*-topology is metrizable.

Proof: Let $\{x_k : k = 1,2,\ldots\}$ be a dense subset of V and define a metric d on $V°$ by

$$d(L_1,L_2) = \sum_{k=1}^{\infty} 2^{-k}|L_1(x_k) - L_2(x_k)|$$

because $|L_j(x_k)| \le 1$ for each k and $j = 1,2$, d is well-defined. Now suppose N is a weak* neighborhood of $L_1 \in V°$. We want to show that N contains a set of the form $\{L \in V° : d(L_1,L) < \delta\}$. We may suppose

$$N = \{L \in V° : |L(y_j) - L_1(y_j)| < \varepsilon,\ j = 1,2,\ldots,n\}$$

where $y_j \in V$. Because $\{x_k\}$ is dense in V we can find, for each $j = 1,2,\ldots,n$, integers $k(j)$ such that $y_j - x_{k(j)} \in (\varepsilon/4)V$. This implies $|L(y_j - x_{k(j)}) - L_1(y_j - x_{k(j)})| < \varepsilon/2$. Thus N contains

$\{L \in V^\circ : |L(x_{k(j)}) - L_1(x_{k(j)})| < \varepsilon/2,\ j = 1,2,\ldots,n\}$. Let $k_0 = \max\{k(1),k(2),\ldots,k(n)\}$ and let $\delta = 2^{-k_0-1}\varepsilon$. Then $d(L_1,L) < \delta$ implies $|L(x_{k(j)}) - L_1(x_{k(j)})| < \varepsilon/2$ and so $L \in N$.

Thus the d-topology on V° is larger than the weak*-topology. To get the other direction it is clear that the d-neighborhood

$$\{L \in V^\circ : d(L_1,L) < \varepsilon\}$$

contains the weak*-neighborhood

$$\{L \in V^\circ : |L(x_k) - L_1(x_k)| < \varepsilon/2^{k+1},\ k = 1,2,\ldots,n\}$$

provided n is so large that $\sum_{n+1}^{\infty} 2^{-k+1} < \varepsilon/2$. QED

The importance of (17.18) relative to our goal (17.16) is that, once a topological question on E^* is reduced to a corresponding question on V°, (17.18) shows us that sequences are adequate.

Define a topology on E^*, called the bw*-topology, as follows: a set $M \subset E^*$ is closed if and only if $M \cap V^\circ$ is weak*-closed for every neighborhood V of 0 in E.

(17.19) Lemma. *Let E be a Fréchet space. Then any neighborhood of 0 in the bw*-topology of E^* contains a set of the form A° where A is a sequence $\{x_n\}$ with $x_n \to 0$ in E.*

Proof: Let $\{p_n : n = 1,2,\ldots\}$ be an increasing sequence of seminorms on E which determine its topology (see the discussion after 17.9) and let $V_n = \{x \in E : p_n(x) < 1/n\}$, $n = 1,2,3,\ldots$. Every neighbor hood V of 0 in E contains some V_n and so $V^\circ \subset V_n^\circ$. Thus a neighborhood W of 0 in the bw*-topology is characterized by $W \cap V_n^\circ$ being a neighborhood of zero in the relative weak*-topology on V_n° for every n. Let $n = 1$. By definition of bw^*, there is a finite set $A_1 \subset E$ such that $A_1^\circ \cap V_1^\circ \subset W$. We show by induction that there exist finite sets A_n, $n = 2,3,\ldots$ such that $A_n \subset V_{n-1}$ and such that $(A_1 \cup\cdots\cup A_n)^\circ \cap V_n^\circ \subset W$ for every n. Suppose this is known for $n = k$ but is not true for $n = k + 1$. Then $(\bigcup_{n=1}^{k} A_n)^\circ \cap V_k^\circ \subseteq W$ but for every finite set $B \subset V_k$ we find $(\bigcup_{n=1}^{k} A_n \cup B)^\circ \cap V_{k+1}^\circ \setminus W$ is non-empty. Call this set K_B and observe that $K_{B_1} \cap K_{B_2} = K_{B_1 \cup B_2}$. Thus the family $\{K_B : B \subset V_k,\ B \text{ finite}\}$ has the finite intersection property. Since each K_B is compact (shown below) in the weak*-topology, it follows that $\bigcap_{B \subseteq V_k} K_B$ is non-empty. But this intersection is just $(\bigcup_{n=1}^{k} A_n \cup V_k)^\circ \cap V^\circ_{k+1} \setminus W = (\bigcup_{n=1}^{k} A_n)^\circ \cap V_k^\circ \setminus W = \emptyset$. Therefore A_{k+1} exists. The sequence $A = \bigcup_{n=1}^{\infty} A_n$ so produced tends to 0 in E since only finitely many elements of A lie outside each V_k. Moreover, $A^\circ \cap V_k^\circ \subset W$ for each k and, since $\cup V_k^\circ = E^*$ we have $A^\circ \subset W$. QED

(17.20) Lemma. *If V is a neighborhood of 0 in a topological vector space E, then V° is compact in the weak*-topology.*

Proof: By Tychonoff's Theorem any infinite product $\prod_{i \in I} X_i$ is compact if each X_i is compact. We embed V° as a closed subspace of such a product where each $X_i = \overline{D(0,1)}$, the closed unit disk in $\mathbb{C}$. The index set we use is V and we define $\Phi : V^\circ \to \prod_{x \in V} \overline{D(0,1)}$ by $\Phi(L) = (L(x))_{x \in V}$. Φ is clearly one-to-one because if $L_1(x) = L_2(x)$ for all $x \in V$ then $L_1(x) = L_2(x)$ for all $x \in E$ and so $L_1 = L_2$. That Φ takes V° into $\prod \overline{D(0,1)}$ follows from the definition of V°. The relative product space topology on $\Phi(V^\circ)$ has the neighborhoods of a point $\Phi(L_0)$ of the form

$$\{\Phi(L) : |L(x_j) - L_0(x_j)| < \varepsilon,\ j = 1,2,\ldots,n\}$$

which correspond exactly to weak*-neighborhoods of L_0 under Φ. That $\Phi(V^\circ)$ is closed is left as an exercise. Since V° is homeomorphic to a closed subspace of a compact space, it is compact. QED

The sets K_B in the proof of (17.19) are closed subsets of V_k°, hence compact.

(17.21) Theorem (Banach-Dieudonné). *If E is a Fréchet space and φ is a continuous linear functional on E^* in the bw*-topology, then φ is continuous in the weak*-topology, and conversely.*

Proof: The weak*-topology is smaller than the bw*-topology so the converse is clear.

Now suppose φ is bw*-continuous. Then $\{L \in E^* : |\varphi(L)| < 1\}$ is open and so contains a set A° where $A = \{x_n : n = 1,2,\ldots\} \subset E$ and $x_n \to 0$ in E. Thus

(17.22) $\quad |\varphi(L)| < 1$ whenever $|L(x_n)| < 1,\ n = 1,2,\ldots$

If $L(x_n) = 0$ for each n then $|kL(x_n)| = 0 < 1$ implies $|\varphi(L)| < 1/k$. As k is arbitrary we see that $\varphi(L) = 0$ whenever $L(x_n) = 0 \, \forall n$. Let c_0 denote the space $\{\{a_n\}_1^\infty : a_n \in \mathbb{C}, a_n \to 0\}$ and topologize it with the norm $\|\{a_n\}_1^\infty\| = \sup_n |a_n|$. If $L \in E^*$ then $\{L(x_n)\}_1^\infty \subset c_0$. Define $\Psi : E^* \to c_0$ by $\Psi(L) = \{L(x_n)\}_1^\infty$. Define a linear functional γ on $\Psi(E^*)$ by

$$\gamma(\Psi(L)) = \varphi(L).$$

By the argument above, γ is well-defined and inequality (17.22) implies

$$|\gamma(\Psi(L))| \leq \|\Psi(L)\|$$

so γ is continuous. By the Hahn-Banach Theorem γ may be extended to $\tilde{\gamma}$ defined on all of c_0 and continuous. By Lemma 17.23 (below) $\tilde{\gamma}$ has the form $\tilde{\gamma}(\{a_n\}_1^\infty) = \sum_{n=1}^\infty a_n b_n$ where $\{b_n\} \subseteq \mathbb{C}$ satisfies $\sum |b_n| < \infty$. Thus

$$\varphi(L) = \gamma(\Psi(L)) = \tilde{\gamma}(\{L(x_n)\}_1^\infty)$$

$$= \sum_{n=1}^\infty b_n L(x_n) = \lim_k L(\sum_{n=1}^k b_n x_n).$$

Since $p_m(\sum_{n=s}^r b_n x_n) \leq (\sum_{n=s}^r |b_n|) \sup_n p_m x_n$

$$\leq C(\sum_{n=s}^r |b_n|) \to 0$$

we see that $\{\sum_{n=1}^k b_n x_n\}_{k=1}^\infty$ is a Cauchy sequence in E and so it has a limit $x_0 \in E$. From the above

$$\varphi(L) = \lim_k L(\sum_{n=1}^k b_n x_n) = L(x_0).$$

Since the functional $L \longrightarrow L(x_0)$ is weak*-continuous, φ is as well. QED

(17.23) Lemma. The dual of c_0 is the space

$$\ell^1 = \{\{b_n\}_1^\infty : b_n \in \mathbb{C}, \sum |b_n| < +\infty\}.$$

Proof: If $L \in (c_0)^*$, let $b_k = L(e^k)$ where e^k is the sequence

$\{\delta_n^k\}_{n=1}^{\infty}$ defined by $\delta_n^k = 1$ if $n = k$, $\delta_n^k = 0$ if $n \neq k$. Because L is continuous we have $|L(\{a_n\})| \leq C \sup_n |a_n|$ for some finite $C \geq 0$. Define a sequence $a^k = \{a_n^k\}_{n=1}^{\infty}$ by

$$a_n^k = \begin{cases} \dfrac{\bar{b}_n}{|b_n|} & \text{if } n \leq k \text{ and } b_n \neq 0 \\ 0 & \text{otherwise} \end{cases}$$

Then $a^k \in c_0$ and $\|a^k\| = 1$. Then

$$\sum_{n=1}^{k} |b_n| = \sum_{n=1}^{k} b_n a_n^k = \sum_{n=1}^{k} L(e^n) a_n^k$$

$$= L\left(\sum_{n=1}^{k} e^n a_n^k\right) = |L(a^k)| \leq C$$

Thus $\sum_{n=1}^{\infty} |b_n| \leq C < +\infty$, and it follows by continuity that $L(\{a_n\}) = \sum a_n b_n$ for arbitrary $\{a_n\} \in c_0$.

It is a trivial exercise to show that every sequence $\{b_n\}$ in ℓ^1 defines a functional L on c_0 satisfying $|L(a)| \leq (\sum |b_n|) \sup_n |a_n|$. QED

The following corollary to (17.21) will suffice to obtain our goal of (17.16).

(17.24) Corollary. Let E be a Fréchet space. A linear subspace M of E^* is weak*-closed if and only if $M \cap V°$ is closed for every neighborhood V of 0 in E.

Proof: It follows easily from Corollary 7.2 of the Hahn-Banach Theorem that a subspace M of E^* is closed in some locally convex topology Ω if and only if for every $L \notin M$ there is an Ω-continuous linear functional φ such that $\varphi|_M = 0$ and $\varphi(L) \neq 0$. Since the bw*-continuous functionals coincide with the weak*-continuous functionals the bw*-closed subspaces coincide with the weak*-closed subspaces. QED

To prove Theorem 17.16 it is enough to show that if M is sequentially closed then $M \cap V°$ is closed for every neighborhood of zero in E. But, by 17.18, $V°$ is metrizable. If $\{L_n\}$ is a sequence in

$M \cap V^\circ$ which converges to a point L, then $L \in M$ if M is sequentially closed and $L \in V^\circ$ since V° is weak*-closed. Thus, if M is sequentially closed, $M \cap V^\circ$ is closed and the theorem is proven.

We close this chapter on an entirely different note: Dirichlet's Theorem on approximation of irrationals by rationals (for use in Chapter 19).

(17.25) Theorem (Dirichlet). If x is an irrational real number then there exist infinitely many rationals p/q in lowest terms such that $|x - p/q| < 1/q^2$.

Proof: Without loss of generality we may suppose $0 < x < 1$. Let a_1 be the last positive integer with $a_1 < 1/x$. Let a_2 be the last positive integer with $a_2 < \dfrac{1}{\frac{1}{x} - a_1}$. Let $\varepsilon_1 = \frac{1}{x} - a_1$, so that $\frac{1}{\varepsilon_1} - 1 < a_2 < \frac{1}{\varepsilon_1}$. Let $\varepsilon_2 = \frac{1}{\varepsilon_1} - a_2$. Since $0 < \varepsilon_1 < 1$ we see that $a_2 \geq 1$ and $0 < \varepsilon_2 < 1$. Define inductively a_{n+1} as the largest integer less than $1/\varepsilon_n$ and $\varepsilon_{n+1} = 1/\varepsilon_n - a_{n+1}$. Now define $p_1 = 1$, $p_2 = a_2$, $q_1 = a_1$, $q_2 = a_1 a_2 + 1$ and inductively

$$p_{n+1} = a_{n+1} p_n + p_{n-1}$$

$$q_{n+1} = a_{n+1} q_n + q_{n-1}.$$

We claim

$$(17.26) \quad \begin{cases} \dfrac{p_1}{q_1} > \dfrac{p_3}{q_3} > \cdots > \dfrac{p_{2n-1}}{q_{2n-1}} > x > \dfrac{p_{2n}}{q_{2n}} > \cdots > \dfrac{p_2}{q_2} \\[2ex] \text{and} \quad \dfrac{p_{2n+1}}{q_{2n+1}} - \dfrac{p_{2n}}{q_{2n}} = \dfrac{1}{q_{2n+1} q_{2n}} < \dfrac{1}{q_{2n}^2}. \end{cases}$$

This will prove the theorem because then

$$\left|x - \frac{p_{2n}}{q_{2n}}\right| < \frac{p_{2n+1}}{q_{2n+1}} - \frac{p_{2n}}{q_{2n}} < \frac{1}{q_{2n}^2}.$$

If $p_{2n}/q_{2n} = p/q$ in lowest terms then $|x - p/q| < 1/q^2$ since $q^2 \leq q_{2n}^2$. Clearly there must be infinitely many such p/q since $q_{2n} \to \infty$ so the estimate $|x - p/q|$ must get arbitrarily small.

To get (17.26) begin by computing

$$p_{2n+1}q_{2n} - q_{2n+1}p_{2n} = p_{2n-1}q_{2n} - q_{2n-1}p_{2n}$$

$$= p_{2n-1}q_{2n-2} - q_{2n-1}p_{2n-2}$$

using the definitions of p_n and q_n.

Consequently

$$\frac{p_{2n+1}}{q_{2n+1}} - \frac{p_{2n}}{q_{2n}} = \frac{1}{q_{2n+1}q_{2n}} (p_{2n+1}q_{2n} - p_{2n}q_{2n+1})$$

$$= \frac{1}{q_{2n+1}q_{2n}} (p_1q_2 - q_1p_2)$$

$$= \frac{1}{q_{2n+1}q_{2n}} .$$

This gives the second half of (17.26). Now $\frac{p_1}{q_1} > x > \frac{p_2}{q_2}$ follows from the definitions. Suppose (17.26) is valid for $n = k$ and consider

$$\frac{p_{2k+2}}{q_{2k+2}} - \frac{p_{2k}}{q_{2k}} = \frac{1}{q_{2k}q_{2k+2}} (p_{2k+2}q_{2k} - p_{2k}q_{2k+2})$$

$$= \frac{a_{2k+2}}{q_{2k}q_{2k+2}} (p_{2k+1}q_{2k} - p_{2k}q_{2k+1}) > 0$$

because of the second part. Thus the even quotients increase and similarly the odds decrease.

Now we show that $p_{2n}/q_{2n} < x < p_{2n-1}/q_{2n-1}$. Suppose this inequality is valid for $n = k$. We wish to show that

$$(17.27) \qquad \frac{p_{2k+2}}{q_{2k+2}} < x < \frac{p_{2k+1}}{q_{2k+1}} .$$

Now $p_{2k+1} = a_{2k+1}p_{2k} + p_{2k-1}$ with a similar equation for q_{2k+1}. Thus the second inequality in (17.27) is equivalent to

$$a_{2k+1}(xq_{2k} - p_{2k}) < p_{2k-1} - xq_{2k-1}.$$

The coefficient of a_{2k+1} is positive so this is equivalent to

$$a_{2k+1} < \frac{p_{2k-1} - xq_{2k-1}}{xq_{2k} - p_{2k}} .$$

This will be valid if we can show the right-hand side is $1/\varepsilon_{2k}$. This is simple if $k = 1$ and, if we let $\gamma_n = \dfrac{xq_n - p_n}{p_{n-1} - xq_{n-1}}$, it is easy to see that $\gamma_{n+1} = \dfrac{1}{\gamma_n} - a_{n+1}$, the same recursion that defines ε_n. Thus $\gamma_n = \varepsilon_n$ for all $n \geq 2$. Thus (17.27) is proved (the left inequality goes the same way). QED

NOTES: The proof of (17.25) is modelled on the theory of continued fractions. That is, any irrational number x (0,1) can be expressed

$$x = \lim_{n\to\infty} \cfrac{1}{a_1 + \cfrac{1}{a_2 + \ddots + \cfrac{1}{a_n}}} = \lim_{n\to\infty} \frac{p_n}{q_n}$$

where the a_i are positive integers chosen as in the proof of (17.25) and p_n/q_n is the nth term expressed as a quotient of integers. See [S.Ya. Khinchin].

Exercises

1. By Proposition 17.8 any norm on $\mathbb{C}^n$ induces the same topology on $\mathbb{C}^n$ as any other norm. Verify this directly for the following norms

$$\|(z_1,z_2,\ldots,z_n)\|_2 = \sqrt{\sum_{k=1}^{n} |z_k|^2}$$

$$\|(z_1,\ldots,z_n)\|_1 = \sum_{k=1}^{n} |z_k|$$

$$\|(z_1,\ldots,z_n)\|_\infty = \max_{1\leq k\leq n} |z_k|$$

2. Use (17.11) to prove that to any weak*-convergent sequence $\{\mu_n\} \subseteq C(G)^*$ there is associated a compact set $K \subseteq G$ such that each μ_n is supported by K. (As usual $G \subseteq \mathbb{C}$ is an open set.)

3. Prove the converse of Theorem 17.13: If K is a compact subset of G and if $\{F_n\}$ is a sequence of germs in $H_0(\mathbb{C}^\wedge\setminus G)$ such that each $F_n(z)$ has an analytic extension to $\mathbb{C}^\wedge\setminus K$ which satisfies $F_n \to 0$ uniformly on compact subsets of $\mathbb{C}^\wedge\setminus K$, then the corresponding functionals L_{F_n} converges weak* to zero in $H(G)^*$.

4. Show that $C(G)$ is a separable Fréchet space.

5. By Lemma 17.23, $c_0^* = \ell^1$. Show that a sequence $b^{(n)} = \{b_k^{(n)}\}_{k=1}^{\infty}$ of elements of ℓ^1 converges to 0 weak* if and only if $\|b^{(n)}\|_1 \equiv \sum_{k=1}^{\infty} |b_k^{(n)}|$ is bounded and $\lim_{n\to\infty} b_k^{(n)} = 0$ for each $k = 1,2,\ldots$ (For necessity use the Principle of Uniform boundedness. The sufficiency is straightforward calculation.)

6. Let $B = \{b = \{b_k\}_{k=1}^{\infty} : \sum |b_k| \le 1\}$. Show in two ways (using Lemma 17.20 and arguing directly from Exercise 5) that any sequence in B has a weak*-convergent subsequence (as elements of $c_0^* = \ell^1$).

7. Show that $M \subseteq H(G)^*$ is weak*-closed if and only if for every compact set $K \subseteq G$ and every sequence $\{L_n\}$ in M supported by K such that the corresponding germs $\{F_n(z)\}$ converge uniformly on compact subsets of $\mathbb{C}^\wedge \setminus K$ to $F(z)$, it follows that F corresponds to an element L_F in M.

8. Show that (17.25) characterizes irrational numbers, that is, if r and s are integers with $(r,s) = 1$, then if p and q are integers with $(p,q) = 1$ and $q > s$, then

$$\left|\frac{r}{s} - \frac{p}{q}\right| > \frac{1}{q^2}$$

(Hint: $\left|\frac{r}{s} - \frac{p}{q}\right| \ge \frac{1}{qs}$.)

§18. Interpolation

In this chapter we study we study interpolation by holomorphic functions from a different point of view from that used in 12. In particular, the Germay Theorem (12.14) says that there exists an entire function that interpolates given values at a given sequence of points. There are three major approaches that can be taken to proving such a theorem. The first is via the Mittag-Leffler Theorem and Weierstrass products, which involves writing an explicit formula. The second is via solving infinitely many linear equations in infinitely many unknowns, the Taylor coefficients. (See [M. Eidelheit] and [P. J. Davis].) The third is via functional analysis--specifically the Banach-Dieudonné theorem. Here we take the third route, obtaining in the process a functional analysis proof of Theorem 12.18.

We note that in the sense of definition (17.9), $H(G)$ is a separable Fréchet space. The separability follows from Runge's Theorem: linear combinations with rational coefficients of functions of the form $(z - a)^{-k}$ and z^k, as k runs through the non-negative integers and a runs through a dense subset of $\mathbb{C}\setminus G$, are dense in $H(G)$. We make use of the results of §17 applied to $E = H(G)$ as well as the concepts that follow.

Let E be a separable Fréchet space and let E^* be its dual space.

(18.1) Definition. A sequence $\{L_n\}$ of continuous linear functionals is said to be interpolating if for every sequence $\{A_n\}$ of complex numbers, there exists an element $f \in E$ such that $L_n(f) = A_n$ for $n = 1,2,3,...$

(18.2) Definition. The sequence $\{L_n\}$ is totally linearly independent if, first of all, it is linearly independent and second, on letting V_n be the linear span of $L_1, L_2,...,L_n$, then if $T_n \in V_n$, $n = 1,2,3,...$ and $T_n \to 0$ weak* as $n \to \infty$, then there exists an N such that $T_n \in V_N$ for $n = 1,2,3,...$

We point out that if a linearly independent sequence L_n is totally linearly independent, then its linear span V is weak*-closed: By (17.16) it suffices to prove V is sequentially closed. If S_n is a sequence in $V = \bigcup_k V_k$ then $S_n \in V_{k_n}$ where k_n is an increasing sequence of integers. Define a sequence with $T_k \in V_k$ by $T_k = S_n$ for $k_n \le k < k_{n+1}$. If S_n converges to S weak*, so does T_k. By the definition (since $\frac{1}{k} T_k \to 0$ weak*) there is an N such that

$T_k \in V_N$ for every k. Since the finite dimensional space V_N is closed, $S \in V_N \subset V$.

(18.3) Theorem. *The sequence $\{L_n\}$ is interpolating if and only if it is totally linearly independent.*

(18.4) Corollary. *If $\{L_n\}$ is an interpolating sequence and if L is linearly independent of the $\{L_n\}$, then $L, L_1, L_2, \ldots$ is an interpolating sequence also.*

(18.5) Definition. *$\{L_n\}$ is said to interpolate rapidly growing sequences if for every sequence of positive constants $\{M_n\}$, there exists an $f \in E$ with $L_1(f) \neq 0$ and $|L_n(f)| \geq M_n |L_{n-1}(f)|$ for $n = 2,3,\ldots$*

(18.6) Corollary. *If $\{L_n\}$ interpolates rapidly growing sequences, then it interpolates all sequences.*

Corollary (18.6) follows from the proof of the theorem because we actually prove that if $\{L_n\}$ interpolates rapidly growing sequences then $\{L_n\}$ is totally linearly independent. It is trivial to verify that if $\{L_n\}$ is interpolating then it interpolates rapidly growing sequences: simply interpolate $\{A_n\}$ where $A_n = M_1 M_2 \cdots M_n$.

Corollary 18.4 is an exercise.

Before we prove Theorem (18.3), let us present the main application. We use the duality theorem for $H(G)$ (9.12) which associates with each L in $H(G)^*$ a germ of an analytic function $[L\hat{}]$ on $\mathbb{C}\hat{}\setminus G$, i.e. $[L\hat{}] \in H_0(\mathbb{C} \setminus G)$. If L is supported by the compact set $K \subset G$ then $L\hat{}$ is defined by $L\hat{}(w) = \tilde{L}(\frac{1}{z - w})$ where $\tilde{L}$ is any extension of L to a linear functional on $C(K)$. $L\hat{}(w)$ is not uniquely determined by L but $[L\hat{}]$ is.

(18.7) Theorem. *The sequence $\{L_n\}$ in $H(G)^*$ is interpolating if and only if*

(a) *it is linearly independent, and*

(b) *for every compact set $K \subset G$, there is an integer $N(K)$ such that if $T_n \in V_n$ and T_n has an analytic continuation to the complement of K, then $T_n \in V_{N(K)}$.*

Proof: We must show that (a) and (b) are equivalent to total linear independence. Suppose first that (a) and (b) are satisfied. Let

$T_n \in V_n$ and $T_n \to 0$ weak*. By Theorem (17.13) there is a compact set K such that $T_n^\wedge$ has an analytic extension to $\hat{\mathbb{C}} \setminus K$. By (b), $T_n \in V_N$ for all n if we choose $N = N(K)$. Thus $\{L_n\}$ is totally linearly independent.

To prove the converse we suppose $\{L_n\}$ is totally linearly independent but that (b) is not valid. Then there exists a compact set K in G such that for any integer N there is a sequence $\{T_k^N\}$ with $T_k^N \in V_k$ and $(T_k^N)^\wedge$ analytic in $\hat{\mathbb{C}} \setminus K$ such that some T_k^N, say $T_{k(N)}^N$, is not in V_N. Thus we obtain a sequence $S_N = T_{k(N)}^N$ such that $S_N^\wedge$ is analytic off K and $S_N \in V_{k(N)} \setminus V_N$. By extracting a subsequence we may suppose $k(N+1) > k(N)$. Define $R_k = S_N$ if $k(N) \leq k < k(N+1)$. Then $R_k \in V_k$, $R_k^\wedge$ is analytic off K, but $\{R_k\}$ is not contained in any one V_N. It follows from (17.13) that $\varepsilon_k R_k \to 0$ weak* for some choice of positive numbers ε_k, but $\{\varepsilon_k R_k\}$ is not contained in any one V_N, contradicting the total linear independence of $\{L_n\}$. QED

Theorem 18.7 easily yields Theorem 12.18 as a corollary:

(18.8) Corollary. Let G be an open set in $\mathbb{C}$ and let z_n, $n = 1,2,3,\ldots$ be an admissible sequence in G (i.e. $z_n \to \partial G$). Let $p_1, p_2, p_3, \ldots$ be a sequence of positive integers. Then given any family of complex numbers $\{a_{n,k} : k = 0,1,\ldots,p_n - 1, n = 1,2,3,\ldots\}$, there exists a function $f \in H(G)$ such that $f^{(k)}(z_n) = a_{n,k}$ for all n and k.

Proof: Let $L_{n,k} \in H(G)^*$ be defined by $L_{n,k}(f) = f^{(k)}(z_n)$. Order these functionals $L_{1,0}; L_{1,1}; \ldots L_{1,p_n-1}, L_{2,0}, \ldots$. Observe that $(L_{n,k})^\wedge(z) = \dfrac{k!}{(z_n - z)^{k+1}}$ so that (a) and (b) of Theorem (18.7) are easy to verify. QED

(18.9) Proof of Theorem 18.3.

Suppose that $\{L_n\}$ is totally linearly independent. Let V be the linear span of $\{L_n : n = 1,2,\ldots\}$ so that $V = \bigcup_n V_n$. We have already remarked that V is weak*-closed. Let $\{A_n\}$ be any sequence of complex numbers and define $\Lambda : V \to \mathbb{C}$ by

$$\Lambda(\sum a_n L_n) = \sum a_n A_n.$$

By the linear independence of the L_n, Λ is well defined. We will show that Λ is weak*-continuous on V and then use the Hahn-Banach

Theorem to extend Λ to a continuous linear function on all of E^* (in the weak*-topology). Since the dual of E^* is again E (17.4), the extended functional corresponds to an element $f \in E$. Hence $L_n(f) = \Lambda(L_n) = A_n$, as desired.

To see that Λ is weak*-continuous on V we need only show its kernel is closed (Theorem 17.6). This is proved exactly the same way as the fact that V is closed: Let $T_n \in \ker \Lambda$ and $T_n \to T$ weak*. We may suppose $T_n \in V_n$ and deduce as before that there is an integer N such that $T_n \in V_N$ for all n. Now $\ker \Lambda \cap V_N$ is closed because it is finite dimensional, and so $T \in \ker \Lambda$. Thus $\ker \Lambda$ is sequentially closed and (17.16) implies it is closed.

For the converse, we suppose only that $\{L_n\}$ interpolates rapidly growing sequences. Then $\{L_n\}$ must be linearly independent. For if $L_n = \sum_{i=1}^{n-1} a_i L_i$ then $|L_n(f)| < \sum_{i=1}^{n-1} |a_i||L_i(f)|$ and simultaneously $|L_n(f)| > M_n M_{n-1} \cdots M_{k+1}|L_k(f)|$. It is clear that a contradiction is obtained if the M_j are chosen with $M_n M_{n-1} \cdots M_{k+1} > n|a_k|$. Now suppose $\{L_n\}$ is not totally linearly independent. Then there would exist $T_k = \sum a_n^{(k)} L_n$ such that the coefficient of highest index is $a_{n_k}^{(k)} \neq 0$ with $n_k \to \infty$, and such that $T_k(f) \to 0$ for each $f \in E$. In particular, for each $f \in E$ there is an integer $k(f)$ such that

$$|\sum a_n^{(k)} L_n(f)| \leq 1 \quad \text{for} \quad k \geq k(f).$$

Thus

$$|a_{n_k}^{(k)} L_{n_k}(f) + \sum_{i=1}^{n_k-1} a_i^{(k)} L_i(f)| \leq 1 \quad \text{for large} \quad k$$

so that

$$|L_{n_k}(f)| \leq \frac{1}{|a_{n_k}^{(k)}|} [1 + \sum_{i=1}^{n_k-1} |a_i^{(k)}||L_i(f)|].$$

Choose $p_k = \dfrac{1}{|a_{n_k}^{(k)}|} \max\{1, |a_i^{(k)}| : i = 1,2,\ldots,n_k - 1\}$. Then $|L_{n_k}(f)| \leq n_k p_k \max\{1, |L_{n_k-1}(f)|, \ldots, |L_1(f)|\}$ for $k \geq k(f)$. Now choose M_n so that $M_n > n$ and $M_{n_k} > n_k^2 p_k$ for $k = 2,3,4,\ldots$ and choose f so that $|L_1(f)| > 1$ and $|L_n(f)| \geq M_n|L_{n-1}(f)|$ for $n = 2,3,4,\ldots$. In particular $|L_1(f)| \leq |L_2(f)| \leq \ldots$ so that $|L_{n_k}(f)| \leq n_k p_k |L_{n_k-1}(f)|$ for $k \geq k(f)$, and yet

$|L_{n_k}(f)| \geq n_k^2 p_k |L_{n_k-1}(f)|$ for all k. This is a contradiction and the theorem is proved.

NOTES: The proofs of Theorem 18.3 and its corollaries and of Theorem 18.7 appear in [P.M. Gauthier and L.A. Rubel]. See also [L.A. Rubel and B.A. Taylor].

Exercises

1. Prove Corollary 18.4: Show that there exists an element $g \in E$ such that $L_n(g) = 0$, $n = 1,2,3,\ldots$ and $L(g) = 1$ (use the Hahn-Banach Theorem and the fact that the span of $\{L_n : n = 1,2,\ldots\}$ is weak*-closed). Then if f satisfies $L_n(f) = A_n$, then $f_0 \equiv f + (B - L(f))g$ satisfies $L_n(f_0) = A_n$ and $L(f_0) = B$.

2. Prove the Mittag-Leffler Theorem 12.17 from Theorem 18.8. (Hint: To obtain a meromorphic function with prescribed principal parts take $f(z) = g(z)/h(z)$ where g and h interpolate the proper sequence of values. For example, if we want near zero

$$(*) \quad \frac{g(z)}{h(z)} = \frac{a_0}{z^\ell} + \frac{a_1}{z^{\ell-1}} + \cdots + \frac{a_{\ell-1}}{z} + \varphi(z)$$

then h must satisfy $h^{(k)}(0) = 0$ for $k = 0,1,\ldots,\ell - 1$ and $h^{(\ell)}(z_n) \neq 0$. Once h has been chosen, (*) determines what values of $g^{(k)}$ to interpolate.

3. In $\mathbb{C}$, let I_n be the closed interval $\{2n \leq x \leq 2n + 1\}$, and let

$$L_n(f) = \int_{I_n} f = \int_{2n}^{2n+1} f(x)dx.$$

Is $\{L_n\}$ an interpolating sequence for $H(\mathbb{C})$?

4. With I_n as in Exercise 3, let $G = \mathbb{C}\setminus \cup I_n$. For each n let Γ_n be a nice curve in G that has winding number 1 around each point of I_n and winding number zero around each I_k, $k \neq n$. Let $L_n(f) = \int_{\Gamma_n} f(z)dz$ for $f \in H(G)$. Is $\{L_n\}$ interpolating?

5. Consider how Theorem 18.3 would change if we asked only that certain sequences A_n could be interpolated. In particular, prove the following Let $\{L_n\}$ be a sequence in E^* which satisfies the property

(*) If $T_k \in \text{span}\{L_n\}$, $T_k = \sum_n a_n^{(k)} L_n$, $k = 1,2,3,\ldots$ such that $T_k \to 0$ weak* then $\sum_n |a_n^{(k)}| \to 0$ as $k \to \infty$.

Assume that $\text{span}\{L_n\}$ is weak*-closed. Then for every bounded sequence of complex numbers $\{A_n\}$ there is an element $f \in E$ such that $L_n(f) = A_n$ for every n.

§19. Gap-Interpolation Theorems

There are many theorems in classical analysis where gaps play a rôle. We take up now some considerations from [N. Kalton and L. A. Rubel] where gaps and interpolation are mixed. The idea is to take the Germay interpolation situation, where we want $f(z_n) = w_n$, $n = 1,2,3,\ldots$ for some entire function f but now require that f have the form

$$f(z) = \sum_{\lambda \in \Lambda} a_\lambda z^\lambda,$$

where Λ is a given set of positive integers. For certain Λ (like $\Lambda = \mathbb{N}$, the set of all positive integers), this interpolation is always possible--provided we require $|z_n| \to \infty$ (and no $z_n = 0$).

For other Λ (like the even integers) it is easy to cook up sequences $\{z_n\}$ and $\{w_n\}$ for which the interpolation $f(z_n) = w_n$ is not possible. A surprising feature of this gap-interpolation problem is that in certain circumstances (where we assume, say, that there are at most two z_n of any given modulus), the main consideration is diophantine approximation, i.e. how well certain real numbers can be approximated by rational numbers.

We shall use the notation Z for a sequence $Z = \{z_n\}$ of distinct non-zero complex numbers with no finite limit point. We always assume that $|z_n| \le |z_{n+1}|$. We let $W = \{w_n\}$ be any sequence of complex numbers. We first give some relevant definitions and notation and then state the results. The principal one provides a necessary and sufficient condition that Λ works for the given sequence Z, for all W. Then we draw some less opaque and more manangeable concrete conclusions from it. At the end we give the proofs.

We say that Z is terminating if it is indexed by $n = 1,2,\ldots,N$, instead of $n = 1,2,3,\ldots$. We consider sequences $\Lambda = \{\lambda_n\}$ of positive integers with $\lambda_j < \lambda_{j+1}$ for all j.

(19.1) Definition. We consider certain classes Ω of sequences Z:

i) Ω_∞ is the class of all Z.

ii) Ω_f is the class of all terminating Z.

iii) Ω_k is the class of all Z such that if $|z_j| = |z_{j+1}| = \cdots = |z_{j+r-1}|$, then $r \le k$. i.e. at most k elements of Z can have any given modulus.

(19.2) Definition. We say that Λ is interpolating for Z (or Z-interpolating) if for every sequence $W = \{w_n\}$ of complex numbers, there exists an entire function of the form

$$f(z) = \sum_{\lambda\in\Lambda} a_\lambda z^\lambda$$

such that $f(z_n) = w_n$ for all n. We say that Λ is Ω-interpolating if Λ is Z-interpolating for all $Z \in \Omega$.

We use this terminology in place of the more descriptive but cumbersome "Λ is an interpolating sequence of exponents for Z". We note that a slight change of point of view has occurred. In §18 we spoke of a sequence of functionals (points) as interpolating for a space $E(H(G))$. Here we fix the sequence (or a class of sequences) of points and apply the terminology to the sequence of exponents Λ. In effect, in the first case we fix the space and adjust the functionals, but in the second we fix the (class of) functionals and adjust the space of entire functions.

(19.3) Definition. Λ is Z-linearly independent if the condition $\sum_{n=1}^{N} a_n z_n^\lambda = 0$ for all $\lambda \in \Lambda$ implies that $a_n = 0$ for $n = 1,2,\ldots,N$. We say that Λ is asymptotically Z-linearly independent if for every $\rho > 0$ there exists an integer $n(\rho)$ so that if $K \geq 0$ and if $|\sum_{n=1}^{N} a_n z_n^\lambda| \leq K\rho^\lambda$ for all $\lambda \in \Lambda$ then $a_n = 0$ for all $n \geq n(\rho)$.

If Λ is asymptotically Z-linearly independent and if $\sum_{n=1}^{N} a_n z_n^\lambda = 0$, then $a_n = 0$ for $n \geq n(\rho)$. Consequently, if $n(0) = \inf_{\rho>0} n(\rho)$ and $Z_0 = \{z_{n_0+1}, z_{n_0+2}, \ldots\}$ then Λ is Z_0-linearly independent. In general neither of these notions implies the other.

(19.4) Definition. Λ is totally Z-linearly independent if it is both Z-linearly independent and asymptotically Z-linearly independent.

By an arithmetic progression we mean a sequence of the form $A = \{qn + p : n = 1,2,\ldots\}$ where $q > 0$ and $p \geq 0$ are integers.

(19.5) Theorem. Λ is interpolating for Z if and only if it is totally Z-linearly independent.

(19.6) Theorem. Λ is Ω_1-interpolating if and only if Λ is infinite.

(19.7) Theorem. Λ is Ω_f-interpolating if and only if Λ has a non-empty intersection with every arithmetic progression A.

(19.8) Theorem. If

(a) $$\lim_{k\to\infty} \frac{\log \lambda_{k+1}}{\lambda_k} = \infty$$

then Λ is not Ω_2-interpolating.

(19.9) Theorem. If

(b) $$\lim_{k\to\infty} \frac{\log \lambda_{k+2}}{\lambda_k} = 0$$

and $\Lambda \cap A$ is not empty for every arithmetic progression A, then Λ is Ω_2-interpolating.

The gist of Theorems (19.8) and (19.9) is roughly that

$\lambda_k = e^{e^{e^{\cdot^{\cdot^{\cdot^{e}}}}}}$ (k times) is the dividing line between Ω_2-interpolating sequences and non-Ω_2-interpolating sequences.

Proof of Theorem (19.5). We will fix Z and Λ. Let $E = H(\mathbb{C})$, the space of all entire functions in the usual topology of uniform convergence on compact sets, and let E_Λ be the closed span in E of $\{z^\lambda : \lambda \in \Lambda\}$. Let $\pi = \pi_\Lambda$ be the projection onto E_Λ, that is,

$$\pi\Big(\sum_{n=0}^{\infty} a_n z^n\Big) = \sum_{\lambda\in\Lambda} a_\lambda z^\lambda.$$

Let L_n be the linear functional on E composed of projection into E_Λ followed by evaluation at z_n:

$$L_n(f) = \pi(f)(z_n).$$

We shall prove that $\{L_n\}$ is interpolating for $H(\mathbb{C})$ (in the sense of (18.1)) if and only if Λ is totally Z-linearly independent. This will prove the theorem because then given $\{w_n\}$ there exists an $F \in H(\mathbb{C})$ such that $L_n(F) = w_n$ for all n. But if $f = \pi(F)$, then f has the required form $f(z) = \sum a_\lambda z^\lambda$ and satisfies $f(z_n) = L_n(F) = w_n$ for all n.

The Cauchy transform $L_n^\wedge(w) = L_n(\frac{1}{w-z})$ satisfies $L_n^\wedge(w) = \sum_{\lambda\in\Lambda} \frac{z_n^\lambda}{w^{\lambda+1}}$. Letting V_n again be the linear span of $L_1, L_2, \ldots, L_n$ we see from Theorem 18.7 that $\{L_n\}$ is interpolating if and only if i) $\{L_n\}$ is linearly independent and ii) for every compact disk K, there exists an integer $N(K)$ such that if $T_n \in V_n$ and $T_n^\wedge$ is analytic outside K, then $T_n \in V_{N(K)}$. Now a typical element $T_n \in V_n$ has a Cauchy transform of the form

$$T_n^\wedge(z) = \sum_{\lambda\in\Lambda} \frac{d_\lambda}{z^{\lambda+1}}$$

where $d_\lambda = \sum_{m=0}^{n} a_m z_m^\lambda$. Of course, $T_n^\wedge(z)$ is analytic in $|z| > R$ if and only if $R \geq \rho$ where

$$\rho = \limsup_{\lambda\to\infty} |d_\lambda|^{1/\lambda}.$$

The equation $\sum a_n L_n = 0$ is equivalent, on taking Cauchy transforms, to $\sum a_n z_n^\lambda = 0$ for all λ. So (i) holds if and only if Λ is Z-linearly independent.

If K is a compact disk $K = \{z : |z| \leq R\}$ and $T_n \in V_n$ with $T_n^\wedge$ analytic off K, then $R \geq \limsup |d_\lambda|^{1/\lambda}$. This is equivalent to

$$|\sum a_m z_m^\lambda| \leq B\cdot R^\lambda \quad \text{for all } \lambda,$$

for some constant $B > 1$. It now follows easily that (ii) holds if and only if Λ is asymptotically Z-linearly independent. QED

(19.10) <u>Proof of Theorem (19.6)</u>. It is clear that Λ must be infinite if it is to be Ω_1-interpolating. The harder part is to show that if Λ is infinite then it is totally Z-linearly independent for every $Z \in \Omega_1$.

Suppose $\sum_{n=1}^{N} a_n z_n^\lambda = 0$ for every $\lambda \in \Lambda$ where Λ is an infinite set and $\{z_n\}$ satisfies $|z_1| < |z_2| < |z_3| < \cdots$. Divide by z_N^λ to get

$$a_N + a_{N-1}\left(\frac{z_{N-1}}{z_N}\right)^\lambda + \cdots + a_1\left(\frac{z_1}{z_N}\right)^\lambda = 0$$

Let $\lambda \to \infty$ through Λ to see that $a_N = 0$. Repeat the argument to obtain successively $a_{N-1} = a_{N-2} = \cdots = a_1 = 0$. Thus Λ is Z-linearly independent.

Now suppose $|\sum_{n=1}^{N} a_n z_n^{\lambda}| < K\rho^{\lambda}$ for every λ. If we let $n(\rho)$ be the first integer for which $|z_{n(\rho)}| > \rho$, we can repeat almost the same argument:

$$(*) \quad \left|a_N + a_{N-1}\left(\frac{z_{N-1}}{z_N}\right)^{\lambda} + \cdots + a_1\left(\frac{z_1}{z_N}\right)^{\lambda}\right| < K\left(\frac{\rho}{|z_N|}\right)^{\lambda}$$

for every $\lambda \in \Lambda$. If $N > n(\rho)$ then $\rho/|z_N| < 1$ and if $\lambda \to \infty$ in (*) we get $a_N = 0$. This argument can be repeated to get $a_{N-1} = a_{N-2} = \cdots = a_{n(\rho)} = 0$, but will not work on any index less than $n(\rho)$.

(19.11) Proof of Theorem (19.7). First, it is clear that the arithmetic nature of Λ is important. For if Λ consisted of only, say, odd integers, then Λ would not be interpolating for even so simple a Z as $\{-1,1\}$: since $f(-1) = -f(1)$ for f of the form $\sum_{\lambda\in\Lambda} a_\lambda z^{\lambda}$, the sequence $W = \{2,2\}$, say, could not be interpolated. Arguments along the same line can be used to show that if $\Lambda \cap A = \emptyset$ for any arithmetic progression A, then Λ is not Ω_f-interpolating.

For the converse direction, suppose $\Lambda \cap A$ is not empty for every arithmetic progression A. It suffices to prove that Λ is Z-linearly independent for every $Z \in \Omega_f$, because the asymptotic Z-linear independence is vacuously satisfied with terminating sequences.

Let

$$R(\mu) = \sum_{n=1}^{N} a_n z_n^{\mu}.$$

We are trying to show that $a_n = 0$ for $n = 1,2,\ldots,N$ if $R(\mu) = 0$ for all $\mu \in \Lambda$. Now

$$\sum_{n=1}^{N} \frac{a_n}{z - z_n} = \sum_{\mu=0}^{\infty} \frac{R(\mu)}{z^{\mu+1}}$$

as remarked in the proof of (19.5). By the Skolem-Mahler-Lech Theorem (see [C. Lech]), whose proof is beyond the scope of this book, the set $M = \{\mu : R(\mu) = 0\}$ is, except for a finite set, the union M^* of finitely many arithmetic progressions $(qn + p)$. The SML Theorem simply says that if you take an ordinary rational function of z and expand it in powers of z (say by long division of polynomials) and let M be the set of indices of those coefficients that vanish, then M has the structure indicated. (By "except for a finite set" we mean that

$(M\setminus M^*) \cup (M^*\setminus M)$ is finite.) Since $R(\lambda) = 0$ for each $\lambda \in \Lambda$ we see that M intersects every arithmetic progression. It follows easily that M^*, as the finite union of arithmetic progressions, must contain all sufficiently large integers. (The complement in $\mathbb{N}$ of an arithmetic progression A is a finite set plus a finite union of arithmetic progressions. The intersection of arithmetic progressions is an arithmetic progression. It follows from these two facts that $\mathbb{N}\setminus M^*$ is a finite union of arithmetic progressions plus a finite set. Since M^* meets every arithmetic projection, $\mathbb{N}\setminus M^*$ is finite.) Thus $\mathbb{N}\setminus M$ is also finite. Therefore for some finite k

$$\sum_{n=1}^{N} \frac{a_n}{z - z_n} = \sum_{\mu=0}^{k} \frac{R(\mu)}{z^{\mu+1}} .$$

The left side has singularities at each z_n where $a_n \neq 0$. But the right has a singularity at most at 0. Since we assume $|z_n| \neq 0$ we must have $a_n = 0$ for $n = 1,2,\ldots,N$. QED

An interesting feature of the Skolem-Mahler-Lech Theorem is that even though both the hypothesis and the conclusion are quite elementary, the only known proofs depend on p-adic analysis.

(19.12) Proof of Theorem (19.8). We proceed here directly without invoking Theorem (19.5) by constructing a sequence $Z = \{z_n\}$ with $|z_{2n}| = |z_{2n+1}| = n$ for all $n = 1,2,\ldots$ for which Λ is not Z-interpolating. We will choose $z_{2n} = ne^{2\pi i x_n}$ with $0 < x_n < 1$ and $z_{2n+1} = n$ in such a manner that, if $w_n = z_{2n}/z_{2n+1}$, there is a function $\varphi(n)$ such that

(1) $|w_n^\lambda - 1| \leq 2\pi/n^\lambda$ if $\lambda \geq \varphi(n)$ and $\lambda \in \Lambda$.

Suppose (1) holds. Then

$$|z_{2n}^\lambda - z_{2n+1}^\lambda| = n^\lambda |w_n - 1|^\lambda,$$

so it follows that if $f(z) = \sum a_\lambda z^\lambda$, then

$$\begin{aligned} |f(z_{2n}) - f(z_{2n+1})| &\leq \sum_{\lambda\in\Lambda} |a_\lambda| n^\lambda |w_n^\lambda - 1| \\ &= \sum_{\lambda<\varphi(n)} |a_\lambda| n^\lambda |w_n^\lambda - 1| + \sum_{\lambda\geq\varphi(n)} |a_\lambda| 2\pi \\ &\leq Kn^{\varphi(n)} + K' \end{aligned}$$

where $K = \sup\{|a_\lambda| : \lambda \in \Lambda\}$ which is finite since f is entire, and $K' = 2\pi \sum_{\lambda\in\Lambda} |a_\lambda|$, which is finite for the same reason ($|a_\lambda| \le M(r)r^{-\lambda}$ by the Cauchy inequalities). Surely, then, there is no such f for which, say, $f(z_{2n}) = 0$ and $f(z_{2n+1}) = n!n^{\varphi(n)}$.

Let us construct $\{x_n\}$. We will use $r(x)$ to mean the distance from x to the nearest integer. Let us hold $n \ge 2$ fixed. Then we choose an integer $\varphi(n) \ge 2$ such that

$$\frac{\log \lambda_{k+1}}{\lambda_k} > \log 2n \quad \text{for all} \quad \lambda_k \ge \varphi(n).$$

Begin with $\lambda_k = \varphi(n)$ and consider the interval $I = \left(-\frac{1}{2\lambda_k n^{\lambda_k}}, \frac{1}{2\lambda_k n^{\lambda_k}}\right)$.

If j is an integer and $y \in \frac{j}{\lambda_k} + I$, then $r(y\lambda_k) < n^{-\lambda_k}$. Let $I_1^{(n)} = \frac{1}{\lambda_k} + I$. Then $\overline{I_1^{(n)}} \subseteq (0,1)$. We wish to choose $I_2^{(n)}$ in such a manner that

1) $\overline{I_2^{(n)}} \subseteq I_1^{(n)}$ and,

2) if $y \in I_2^{(n)}$ then $r(y\lambda_{k+1}) < n^{-\lambda_{k+1}}$.

Since $\frac{1}{\lambda_{k+1}} < (2n)^{-\lambda_k} < (2\lambda_k)^{-1} n^{-\lambda_k}$ = half the length of $I_1^{(n)}$, there is an integer j so that $\left[\frac{j}{\lambda_{k+1}} - \frac{1}{2\lambda_{k+1} n^{\lambda_{k+1}}}, \frac{j}{\lambda_{k+1}} + \frac{1}{2\lambda_{k+1} n^{\lambda_{k+1}}}\right] \subseteq I_1^{(n)}$. Call this $\overline{I_2^{(n)}}$. If $y \in I_2^{(n)}$ then $|y\lambda_{k+1} - j| < \frac{1}{2n^{\lambda_{k+1}}}$ so $r(y\lambda_{k+1}) < n^{-\lambda_{k+1}}$. Note that $\overline{I_2^{(n)}} \subseteq (0,1)$. Again, because $\frac{1}{\lambda_{k+2}} < (2\lambda_{k+1})^{-1} n^{-\lambda_{k+1}} < \frac{1}{2}$ length of $I_2^{(n)}$ we can obtain $I_3^{(n)}$ with $\overline{I_3^{(n)}} \subseteq I_2^{(n)}$ and $r(y\lambda_{k+2}) < n^{-\lambda_{k+2}}$ for all $y \in I_3^{(n)}$. Thus we obtain a decreasing sequence of intervals $I_1^{(n)} \supseteq I_2^{(n)} \supseteq I_3^{(n)} \supseteq \cdots$. Let $x_n \in \bigcap_m I_m^{(n)}$. Since x_n is in every $I_m^{(n)}$ we have $r(x_n\lambda_k) < n^{-\lambda_k}$ for all $\lambda_k \ge \varphi(n)$. Moreover $\overline{I_2^{(n)}}$ does not contain 0 or 1 so $0 < x_n < 1$. Since $r(x_n\lambda_k) < n^{-\lambda_k}$, it follows that $|e^{2\pi i x_n \lambda_k} - 1| < 2\pi n^{-\lambda_k}$. If $n = 1$, any choice of x_n will do. Since $w_n = e^{2\pi i x_n}$ we have $|w_n^\lambda - 1| < 2\pi n^{-\lambda}$ when $\lambda \ge \varphi(n)$. This finishes the proof.

Before we prove Theorem (19.9), we need a lemma.

(19.13) Lemma. If Λ is not Ω_k-interpolating then either

a) There exists a $Z \in \Omega_k$ and constants $a_1, a_2, \ldots, a_n$ not all zero such that $a_1 z_1^\lambda + \cdots + a_n z_n^\lambda = 0$ for all $\lambda \in \Lambda$, or

b) There exists an ε with $0 < \varepsilon < 1$ and non-zero numbers $\xi_1, \ldots, \xi_n$ with $n \le k$, $|\xi_j| = 1$ all $j = 1,2,\ldots,n$, such that for some a_j not all zero, $|a_1 \xi_1^\lambda + \cdots + a_n \xi_n^\lambda| \le K\varepsilon^\lambda$ for all $\lambda \in \Lambda$.

Proof: (a) This is the case when Λ is not Z-linearly independent for some $Z \in \Omega_k$.

(b) This is the case when Λ is not asymptotically Z-linearly independent. The details are left as an exercise. They are similar to the arguments in (19.10).

(19.14) Proof of Theorem (19.9). By the lemma, if Λ satisfies the hypothesis, but is not Ω_2-interpolating, then there exists an ε with $0 < \varepsilon < 1$ and numbers K', a_1, a_2, z_1, z_2 with $z_1 \neq z_2$ and $|z_1| = |z_2| = 1$ such that

(i) $|a_1 z_1^\lambda + a_2 z_2^\lambda| \le K'\varepsilon^\lambda$ for all $\lambda \in \Lambda$.

(We can exclude the case $n = 1$ in 19.13(b) since that would say $|a_1| < K\varepsilon^\lambda$, an absurdity.) Dividing (*) through by $|a_1 z_2^\lambda|$ we get

(ii) $|w^\lambda - A| < K\varepsilon^\lambda$ for all $\lambda \in \Lambda$

where $w = z_1/z_2$, $A = -a_2/a_1$ and $K = K'/|a_1|$. It is clear, on letting $\lambda \to \infty$, that $|A| = 1$. Writing $A = e^{2\pi i\alpha}$, $w = e^{2\pi i x}$, we can rewrite (ii) as

$$r(\lambda x - \alpha) < K_1\varepsilon^\lambda \quad \text{for all } \lambda \in \Lambda$$

for some constant K_1. Furthermore, x must be irrational. For if we had $x = p/q$, then we could choose λ large and divisible by q (since Λ meets every arithmetic progression) to obtain $r(\alpha) < k_1\varepsilon^\lambda$ for arbitrarily large λ. Thus α would be an integer. Now choosing $\lambda \equiv 1 \bmod q$ we would have $r(x) = r(\lambda x - \alpha) < K_1\varepsilon^\lambda$ for arbitrarily large λ and so x is an integer. This implies $z_1 = z_2$, contrary to hypothesis.

We now appeal to the Dirichlet Theorem on approximation by rationals (Theorem 17.25), to get infinitely many integers q and p such that

$$|x - \frac{p}{q}| < \frac{1}{q^2} \quad \text{and} \quad (p,q) = 1.$$

For such a large q, pick λ_{n+1} in Λ as large as possible so that $4\lambda_{n+1} \le q$. Then $\lambda_{n+2} > q/4$. We have

$$|\lambda_n x - \lambda_n \frac{p}{q}| \le \frac{\lambda_n}{q^2} \le \frac{1}{4q}$$

and

$$|\lambda_{n+1} x - \lambda_{n+1} \frac{p}{q}| \le \frac{1}{4q} .$$

Hence

$$|(\lambda_{n+1} x - \lambda_n x) - (\lambda_{n+1} - \lambda_n) \frac{p}{q}| \le \frac{1}{2q} .$$

But

$$\frac{\lambda_{n+1} - \lambda_n}{q} p \ge \frac{1}{q}$$

since the numerator must be a positive integer. Hence

$$|\lambda_{n+1} x - \lambda_n x| \ge \frac{1}{2q} .$$

But, by the triangle inequality,

$$\begin{aligned} |\lambda_{n+1} x - \lambda_n x| &= r(\lambda_{n+1} x - \lambda_n x) \\ &\le r(\lambda_{n+1} x - \alpha) + r(\lambda_n x - \alpha) \\ &\le 2K_1 \varepsilon^{\lambda_n} \end{aligned}$$

and so

$$2K_1 \varepsilon^{\lambda_n} \ge \frac{1}{2q} > \frac{1}{8\lambda_{n+2}}$$

infinitely often. This implies, on taking logarithms, that

$$\frac{\log \lambda_{n+2}}{\lambda_n} > \log \frac{1}{\varepsilon} - \frac{\log 16K_1}{\lambda_n}$$

infinitely often. This contradicts hypothesis 19.9(b), completing the proof.

NOTES: The results in this chapter were taken from [Kalton and Rubel].

Exercises

1. Prove that if Λ is Ω_f-interpolating and if A is an arithmetic progression, then $\Lambda \cap A \neq \emptyset$.

2. Provide the details of the proof of Lemma (19.13).

3. Prove that Λ is Ω_∞-interpolating if Λ contains, for every positive integer k, a set of consecutive integers with k elements. Outline: (I) Show that if Λ is not Z-linearly independent for some $Z = \{z_n\}$, then $\mathbb{N}$ is not Z'-linearly independent for some Z', a contradiction:

a) $\exists\, a_1, a_2, \ldots, a_n$ and $z_1, z_2, \ldots, z_n$ such that for all k, $a_1 z_1^{\lambda_k + t} + \cdots + a_n z_n^{\lambda_k + t} = 0$ for some $\lambda_k \in \Lambda$, $t = 1, 2, \ldots, k$.

b) Consider the set

$$\left\{ \left(\frac{a_1 z_1^{\lambda_k}}{z_n^{\lambda_k}}, \frac{a_2 z_2^{\lambda_k}}{z_n^{\lambda_k}}, \ldots, a_n \right) : k = 1, 2, \ldots \right\}.$$

If $(b_1, \ldots, b_n)$ is a limit point of this set then $b_1 z_1^t + \cdots + b_n z_n^t = 0$ $\forall\, t \in \mathbb{N}$.

(II) Make a similar argument if Λ is not asymptotically Z-linearly independent.

4. Show that no arithmetic progression can be Ω_f-interpolating. Show that no Λ which is a finite union of arithmetic progressions can be Ω_f-interpolating unless $\mathbb{N} \setminus \Lambda$ is finite.

5. Let $\Omega_{\mathbb{R}}$ denote all sequences $Z = \{z_n\}$ such that each z_n is real. Show that Λ is $\Omega_{\mathbb{R}}$-interpolating if and only if Λ is infinite.

6. Let Ω_+ be the set of all sequences $Z = \{z_n\}$ of distinct points a_n such that whenever $|z_n| = |z_{n+1}|$ then $z_n = -z_{n+1}$ (i.e. for each n, either $|z_{n+1}| > |z_n|$ or $z_{n+1} = -z_n$). Show that Λ is Ω_+-interpolating if and only if Λ contains infinitely many odd numbers and infinitely many even numbers.

§20. First-Order Conformal Invariants

(20.1) The Theme of this section is the following: Suppose you find yourself on a plane domain, with only a restricted logic at your disposal: how closely can you determine which domain you are on--up to conformal equivalence? This leads to a study of a system of conformal invariants, the first-order conformal invariants (FOCI), which are obtained from the elementary properties of the algebra (or ring) of analytic functions on plane domains. Although the formal definition of FOCI is given in the terminology of mathematical logic, these invariants are nonetheless all included within the framework of classical function theory. Each of the FOCI corresponds to an elementary assertion about analytic functions that can be understood without any knowledge of mathematical logic.

For each domain G we let $H(G)$ denote the ring of holomorphic functions on G; also, we let $H_{\mathbb{C}}(G)$ denote $H(G)$ as an algebra over the field $\mathbb{C}$ of complex numbers. By the ring language we mean the first-order formal language appropriate to the structure $H(G)$. This language has basic symbols for addition and multiplication of functions, as well as the usual logical symbols: propositional connectives $\wedge$ ("and"), $\vee$ ("or"), $\neg$ ("not") and $\Rightarrow$ ("implies") as well as quantifier symbols $\forall$ ("for all") and $\exists$ ("there exists") together with variables f, g, h,..., which range over $H(G)$. For convenience, we also include in the ring language a constant symbol which is a name for the constant function $i = \sqrt{-1}$. Formulae and sentences in this language are finite combinations of these basic symbols, arranged according to the obvious formal rules of grammar. A key restriction is that the language is first-order which means that we can only use quantifiers over variables that represent elements of $H(G)$ and not over variables that represent subsets, ideals, relations, etc., in $H(G)$. Roughly speaking, this means we may examine every element in $H(G)$ to see if it satisfies some property, but we may not examine every ideal to see if it has some property of ideals. Thus "$\exists f(f^3 = -f)$" is in the first-order ring language, but "$\exists I$ (I is an ideal and $I \cdot I = I$)" is not. It is required also that the expressions in a first-order language be finite in length.

The algebra language is the first-order language appropriate to the algebra $H_{\mathbb{C}}(G)$. It is formed by adding to the ring language a 1-place predicate Const(·). In $H_{\mathbb{C}}(G)$ we interpret Const(f) to mean that f is a constant function. (Thus we are identifying $\mathbb{C}$ with the field of constant functions in $H(G)$ and are using the multiplication of functions in $H(G)$ to play the rôle of the scalar multiplication in the algebra.)

In dealing with rings and algebras, the expressive power of first-order sentences is reasonably well understood. For example, to say that a ring is commutative is first-order $(\forall x \forall y(xy = yx))$ but, at least superficially (and in fact) to say that a ring is simple is not first-order, since it seems to require quantification over subsets ("there does not exist a proper two-sided ideal").

It is first-order to say that a ring has at least 2 elements $(\exists x \exists y(x \neq y))$ but it is not first-order to say that it is infinite, since this requires a sentence of infinite length, as one might suspect.

Now if G and G' are conformally equivalent, then the algebras $H_{\mathbb{C}}(G)$ and $H_{\mathbb{C}}(G')$ are isomorphic. If $T : G \to G'$ is the one-to-one conformal mapping, then the algebra isomorphism takes $f \in H_{\mathbb{C}}(G')$ to the composition $f \circ T \in H_{\mathbb{C}}(G)$. This implies that for each sentence A in the algebra language, the assertion "A holds in $H_{\mathbb{C}}(G)$" is a conformal invariant of G. That is, if A holds in $H_{\mathbb{C}}(G)$, then A holds in $H_{\mathbb{C}}(G')$ for any G' which is conformally equivalent to G. This can be turned into a numerical invariant by taking 1 to mean A is true and 0 to mean that it is false. It is these invariants of G which are the "first-order conformal invariants" in the title of this section. Note the relevance of Bers' Theorem of §16 and Theorem 5.4. We emphasize that a name for the constant $\sqrt{-1}$ has been included in both the ring language and the algebra language. This is because one cannot generally determine G up to conformal equivalence using $H(G)$ (or $H_{\mathbb{C}}(G)$). For example, the map taking f to $\tilde{f}$ defined by $\tilde{f}(z) = \overline{f(\bar{z})}$ is an isomorphism from $H(G)$ to $H(\bar{G})$: evidently it takes i to $-i$. (Note that Theorem 5.4 is not quite applicable unless a name for $\sqrt{-1}$ is included in the language for $H_{\mathbb{C}}(G)$. While the $\mathrm{Const}(\cdot)$ enables us to tell if a function is constant and an isomorphism of $H_{\mathbb{C}}(G)$ must preserve constants, there is in general no algebraic way of determining whether it sends i to i or to $-i$.) Now for the mathematics itself.

(20.2) To begin with, note that the constant functions 0 and 1 are definable using their first-order properties in $H(G)$ as a ring. Also, the property that f is a unit in $H(G)$ is first-order expressible: $\exists g(fg = 1)$. Thus we can express, for any definable constant c, the condition that f omits the value c on G, by saying that $f - c$ is a unit in $H(G)$. These considerations lead quickly to a first-order characterization of those domains which have exactly n holes, for each particular value of n. For example, G is simply connected if and only if every unit is a square. (Exercise). More generally, consider any domain G and a unit u in $H(G)$. For each curve γ in G set

$$W(u,\gamma) = \frac{1}{2\pi i}\int_{\gamma} \frac{u'(z)}{u(z)}\, dz,$$

the winding number of $u \circ \gamma$ about 0. It is a fact (Exercise) that the unit u is a square in $H(G)$ if and only if $W(u,\gamma)$ is an even integer whenever γ is a simple closed curve in G.

Let $u_1, u_2, \ldots, u_n$ be units in $H(G)$. For each set $S \subseteq \{1,2,\ldots,n\}$ we set

$$u_S = \prod_{j \in S} u_j, \quad u_\emptyset = 1$$

(20.3) Definition. The units $u_1, \ldots, u_n$ are called a system of Johnson units for G if for each unit $f \in H(G)$, there is a unique set $S \subseteq \{1,2,\ldots,n\}$ such that fu_S is a square in $H(G)$.

(20.4) Proposition. The domain G has exactly n holes if and only if there exists a system of n Johnson units for G.

Proof: First suppose G has exactly n holes. Let $z_1, z_2, \ldots, z_n \in \mathbb{C}$ be selected so that z_j lies in the jth hole of G. Set $u_j(z) = z - z_j$ for $z \in \mathbb{C}$, $j = 1,2,\ldots,n$. For each j, let γ_j be a simple closed curve in G which winds once around z_j but does not wind around z_i for $i \neq j$. Given any unit $f \in H(G)$, let $S = \{i : W(f,\gamma_i) \text{ is odd}\}$. Then it is easy to see that fu_S is a square in $H(G)$, since

$$(*) \qquad W(fu_S,\gamma_i) = \begin{cases} W(f,\gamma_i) & \text{if } i \notin S, \text{ and} \\ W(f,\gamma_i) + 1 & \text{if } i \notin S, \end{cases}$$

hence $W(fu_S,\gamma_i)$ is always even. It follows from the Cauchy Theorem (11.5 et seq.) that $W(fu_S,\gamma) = \sum_{j=1}^{n} W(\gamma : z_j) W(fu_S,\gamma_j)$ for any simple closed curve γ in G. This shows that $u_1, \ldots, u_n$ is a system of Johnson units for G (the uniqueness of S is clear from (*)).

Conversely, suppose $u_1, \ldots, u_n$ is any system of Johnson units for G. For each unit f in $H(G)$, let $S = S(f)$ be the unique subset of $\{1,2,\ldots,n\}$ for which $f \cdot u_S$ is a square in $H(G)$. Suppose first that

G has more than n holes. Then we may find points $z_1, z_2, \ldots, z_m (m > n)$ which lie in different holes and let $v_1, v_2, \ldots, v_m$ be the units defined by $v_j(z) = z - z_j$ on G. For each $T \subseteq \{1,2,\ldots,m\}$ we define $T' = S(v_T)$ where $v_T = \prod_{j \in T} v_j$. Then $v_T u_{T'}$ is a square in $H(G)$. Evidently v_T is not a square in $H(G)$ unless $T = \emptyset$. Suppose T_1, T_2 are subsets of $\{1,2,\ldots,m\}$ and $T_1' = T_2'$. It follows that $v_{T_1} v_{T_2}$ is a square and hence that v_T is a square, where $T = (T_1 \setminus T_2) \cup (T_2 \setminus T_1)$. Therefore $T = \emptyset$ and so $T_1 = T_2$. It follows that $T \mapsto T'$ defines a one-to-one mapping from subsets of $\{1,2,\ldots,m\}$ to subsets of $\{1,2,\ldots,n\}$. This contradicts the assumption that $m > n$. Thus G has at most n holes.

Suppose now that G has exactly k holes, $k \leq n$. Then we obtain, as in the first part of the proof, a system of Johnson units $v_1, \ldots, v_k$ for G. But then we obtain a mapping $S \mapsto S'$ from subsets of $\{1,2,\ldots,n\}$ to subsets of $\{1,2,\ldots,k\}$ such that $v_{S'} \cdot u_S$ is a square for each S. If $S_1' = S_2'$ we obtain the result as above that $1 \cdot u_S$ is a square where $S = (S_1 \setminus S_2) \cup (S_2 \setminus S_1)$. By the uniqueness of S we get $S = \emptyset$. Again we get a contradiction unless $k = n$. This completes the proof.

Since the assertion that there exists a system of Johnson units for G can clearly be expressed using a sentence of the first-order ring language we have proved the following result.

(20.5) Corollary. *For each* $n \in \mathbb{N}$ *there is a sentence* S_n *in the ring language such that for any domain* G, S_n *holds in* $H(G)$ *if and only if* G *has exactly* n *holes.*

Thus, having n-holes is first-order expressible. However, having finitely many holes may not be.

We can distinguish between the two types of simply connected domains by a sentence in the ring language which expresses the content of Picard's Little Theorem--which holds for $H(\mathbb{C})$ but not for $H(\mathbb{D})$. We shall not prove this theorem in this book, but we shall show how it can be used.

(20.6) Picard's Little Theorem. *An entire function that omits two distinct values must be constant.*

Let P be the sentence that expresses the following: *If* $f - 1$ *and* $f - i$ *are units, but* f *is not a unit, then* $f = 0$. Then P is true in $H(\mathbb{C})$ but it is false in $H(G)$ when G is conformally equivalent to the unit disk $\mathbb{D}$. In particular, this, together with Exercise 1 (the special case $n = 0$ of Corollary 20.5) gives a first-order characterization of the unit disk $\mathbb{D}$.

We can achieve this without using the Picard Theorem if we allow ourselves to use the algebra language, i.e. pass to $H_{\mathbb{C}}(G)$.

(20.7) Definition. An element f of $H(G)$ is called a prime if f is not a unit and if $f = gh$ implies g is a unit or h is a unit.

It is easy to see that f is a prime if and only if f has exactly one zero on G, counting multiplicity (Exercise).

(20.8) Proposition. A function $f \in H_{\mathbb{C}}(G)$ is univalent if and only if for all $c \in \mathbb{C}$, $f - c$ is either a unit or a prime.

The proof is an exercise.

Now consider the following statement which can be expressed in a first-order way in the language of $H_{\mathbb{C}}(G)$:

S : If f and g are univalent then there exists a constant b such that $bf - g$ is a constant.

S is clearly not true in $H_{\mathbb{C}}(\mathbb{D})$. To show that it enables us to distinguish $\mathbb{D}$ and $\mathbb{C}$ we need to recall Exercise 17 of Chapter 2:

(20.9) Proposition. Every univalent entire function has the form $az + c$ for some constants c and a.

Proof: Let f be univalent. Then $f(\mathbb{D})$ is open, so $f(\mathbb{C}\setminus\mathbb{D}) \subseteq \mathbb{C}\setminus f(\mathbb{D})$ implies that $f(U)$ is not dense in $\mathbb{C}$ for some deleted neighborhood U of ∞. By the Casorati-Weierstrass Theorem (2.25), f has at worst a pole at ∞. That is, f is a polynomial. To be uni-falent f has at most one zero, so it is linear. QED

Proposition 20.9 shows that S is true in $H_{\mathbb{C}}(\mathbb{C})$.

(20.10) Corollary. There exists a sentence R in the language of $H_{\mathbb{C}}(G)$ that holds if and only if G is conformally equivalent to the unit disk.

Proof: See Exercise 5.

NOTES: For a general treatment of mathematical logic see [J. R. Shoenfield]. Not much is known about first-order conformal invariants. For more information see [J. Becker, C. W. Henson and L. A. Rubel]. The reader is welcome to do his own thing.

Exercises

1. Prove that G is simply connected if and only if every unit in $H(G)$ is a square.

2. Show that a unit $u \in H(G)$ is a square if and only if $W(u,\gamma)$ is an even integer for every simple closed curve γ in G. (Hint: If $W(u,\gamma)$ is always even, let $a \in G$ and define

$$g(z) = \exp\left(\frac{1}{2}\int_a^z \frac{u'(\zeta)}{u(\zeta)}\, d\zeta\right).$$

show that g is independent of the path chosen in G from a to z and that g^2u^{-1} is constant.)

3. Show that an element $f \in H(G)$ is prime if and only if f has exactly one zero in G, counting multiplicity.

4. Prove Proposition 20.8.

5. Prove Corollary 20.10. (Hint: Let R say that G is simply connected and S is not true, in the first-order algebra language.)

6. Show that the constant function $i = \sqrt{-1}$ in $H(G)$ is uniquely defined by a property expressible in the first-order ring language.

7. Show that the following properties (or definitions) of a ring (or its elements) are first-order.
   a. multiplication distributes over addition.
   b. The units (resp. non-units) are closed under multiplication.
   c. f divides g.
   d. The ideal generated by $f_1, f_2, \ldots, f_n$.
   e. f is the greatest common divisor of $g_1, g_2, \ldots, g_n$.

8. Show that the statement "f is a bounded function" is equivalent to

a first-order statement in $H_{\mathbb{C}}(\mathbb{C})$. What about $H_{\mathbb{C}}(\mathbb{D})$? What if we added some topological language to $H_{\mathbb{C}}(\mathbb{D})$? (Say by adding the predicate $\lim(\{f_n\},f)$ which is interpreted to mean $\{f_n\}$ is a sequence and f is its limit.)

9. Show that "f has a zero of order 2" is equivalent to a first-order sentence in $H(G)$.

References

J. Becker, C. Ward Henson and L. A. Rubel, First-order conformal invariants, Annals of Math. 112(1980), 123-178.

L. Bers, On rings of analytic functions, Bull. Amer. Math. Soc. 54(1948), 311-315.

C. Carathéodory, [1] Theory of Functions of a Complex Variable, Chelsea Publishing Company, New York, 1954, 2 Volumes.

__________, [2] Conformal Representation, Cambridge University Press, Cambridge, 1958.

D. Challener and L. A. Rubel, A converse to Rouché's Theorem, Amer. Math. Monthly 89(1982), 302-305.

P. J. Davis, Interpolation and Approximation, Blaisdell Publishing Co., New York, 1963.

M. Eidelheit, Zur Theorie des Systeme linearer Gleichungen, Studia Math. 6(1936), 139-148.

J. Dugundji, Topology, Allyn and Bacon, Boston, 1966.

T. Esterman, Complex Numbers and Functions, The Athlone Press, London, 1962.

O. J. Farrell, On approximation by polynomials to a function analytic in a simply connected region, Bull. Amer. Math. Soc. 41(1934), 707-711.

P. Gauthier and L. A. Rubel, [1] Interpolation in separable Fréchet spaces with application to spaces of analytic functions, Can. J. Math. 27(1975), 1110-1113.

__________, [2] Holomorphic functionals on open Riemann surfaces, Can. J. Math. 28(1976), 885-888.

I. Glicksberg, A remark on Rouché's Theorem, Amer. Math. Monthly 83(1976), 186-187.

A. Grothendieck, Sur certain espaces de fonctions holomorphes I, J. für die Reine und Angew. Math. 192(1953), 35-64.

O. Helmer, Divisibility properties of integral functions, Duke Math. J. 6(1940), 345-356.

M. Henriksen, [1] On the ideal structure of the ring of entire functions, Pacific J. Math. 2(1952), 179-184.

__________, [2] On the prime ideals of the ring of entire functions, Pacific J. Math. 3(1953), 711-720.

E. Hille, Analytic Function Theory, Ginn and Co., Boston, Vol. I, 1959; Vol. II, 1962.

L. Hörmander, An Introduction to Complex Analysis in Several Variables, North Holland/American Elsevier, Amsterdam/New York, 1973.

H. Iss'sa, On the meromorphic function field of a Stein variety, Ann. of Math. 83(1966), 34-46.

N. Kalton and L. A. Rubel, Gap-interpolation theorems for entire functions, J. für die Reine und Angew. Math. 316(1980), 71-82.

J. L. Kelley, General Topology, D. Van Nostrand, New York, 1955.

J. L. Kelley, and I. Namioka, Linear Topological Spaces I, D. Van Nostrand, Princeton, N.J., 1963.

A. Ya. Khinchin, Continued Fractions, University of Chicago Press, Chicago, 1964.

G. Köthe, [1] Dualität in der Funktionentheorie, J. für die Reine und Angew. Math. 191(1953), 30-49.

__________, [2] Topological Vector Spaces, Springer-Verlag, New York, 1969.

C. Lech, A note on recurring series, Arkiv för Mat. 2(1952), 417-421.

L. A. Rubel, [1] How to use Runge's Theorem, l'Enseignement Math. 22 (1976), 185-190 with Errata in Vol. 23(1977).

__________, [2] Solution of problem no. 6117 (proposed by M. J. Pelling), Amer. Math. Monthly 85(1978), 505-506.

L. A. Rubel and A. L. Shields, Bounded approximation by polynomials, Acta Math. 112(1964), 145-162.

L. A. Rubel and B. A. Taylor, Functional analysis proofs of some theorems in function theory, Amer. Math. Monthly 76(1969), 483-489.

S. Saks and A. Zygmund, Analytic Functions, Monografie Mathematyczne, Nakladem Polskiego Towarzystwa Matematycznego, Warszawa-Wroclaw 1952.

J. Sebastiao e Silva, [1] As funcoes analyticas e a analise funcional, Thesis, Lisbon 1948, Published in Port. Math. 9(1950).

__________, [2] Sui fundamenti della teoria dei funzionali analitici, Port. Math. 12(1953), 1-46.

W. Seidel and J. L. Walsh, On approximation by Euclidean and non-Euclidean translations of an analytic function, Bull. Amer. Math. Soc. 47 (1941), 916-920.

C. L. da Siva Dias, Espaços Vectoriais topológicos e sua aplicacão nos espacos funcionais analiticas, Thesis, São Paulo 1952 (Separata do Boletin da Sociedade de Matematica de São Paulo, Vol. 5°, fasciculo 1° et 2°, dezembro 1950.

J. R. Shoenfield, Mathematical Logic, Addison-Wesely, Reading, Mass., 1967.

H. G. Tillmann, Dualität in der Funktionentheorie, J. für die Reine und Angew. Math. 195(1956), 76-101.